MW01138172

CHEVROLET SS
MUSCLE CAR
RED BOOK

Peter C. Sessler

MBI Publishing Company

*Special thanks go to all the enthusiasts who
helped with this book, particularly Joe Bertalan,
Susan Bowden, Ed Conrads, Bob Konner, Mark Meekins,
Bill Petriko, Rob Quinn, Len Williamson, National Chevelle
Owners Association, National Monte Carlo Owners Association
and Chevrolet Motor Division.*

First published in 1991 by MBI Publishing Company, PO Box 1,
729 Prospect Avenue, Osceola, WI 54020-0001 USA

The information in this book is true and complete to the best of
our knowledge. All recommendations are made without any
guarantee on the part of the author or Publisher, who also
disclaim any liability incurred in connection with the use of this
data or specific details.

We recognize that some words, model names, and designations, for
example, mentioned herein are the property of the trademark
holder. We use them for identification purposes only. This is not an
official publication.

MBI Publishing Company books are also available at discounts in
bulk quantity for industrial or sales-promotional use. For details
write to Special Sales Manager at Motorbooks International
Wholesalers & Distributors, 729 Prospect Avenue, PO Box 1,
Osceola, WI 54020-0001 USA.

Library of Congress Cataloging-in-Publication Data
Sessler, Peter C.
    Chevrolet SS red book / Peter C. Sessler.
      p. cm.
    ISBN 0-87938-501-4
    1. Chevrolet automobile. 2. Impala automobile. 3. Chevy II
automobile. 4. Nova automobile. I. Title. II. Series.
TL215.C5S47 1991
629.222'2—dc20   90-24614

On the front cover: The 1970 Chevelle Super Sport 454
convertible owned by Alan Aertz of San Jose, California.
Jerry Heasley

Printed in the United States of America

Contents

Introduction

This book is designed to help the Chevrolet enthusiast determine the authenticity and originality of any Super Sport Impala, Chevelle, Camaro or Monte Carlo produced between 1961 and 1972. Each chapter covers a model year in which a Super Sport model or SS Option Package was available. Included are production figures; serial number decoding information; engine codes; carburetor, distributor and cylinder head numbers; color and trim codes; options with pricing; and selected facts. Not all engines are covered in the same detail because of space limitations.

A lot of supercars were built in the sixties, but by far the most popular ones have the familiar SS letters on them. Chevrolet first introduced an SS Option Package on the 1961 Impala. By 1965, the intermediate Chevelle had the right image and performance to attract the youthful buyer—which it did in ever-increasing numbers. The Chevelle SS 396s and SS 454s evoke images of style and power. These intermediates gave the Super Sport the reputation it enjoys even now. By 1973, however, it was clear that not much demand existed for a Super Sport Chevelle.

For the enthusiast, the most important number in any Chevrolet Super Sport is its VIN, or vehicle identification number. From 1961 to 1964, the VIN consisted of eleven digits, which broke down to model year, body style and series, assembly plant and consecutive sequence number. In 1965, it expanded to thirteen digits but did not offer any additional information. The extra two digits were used to identify the model and body series. In 1972, Chevrolet adopted a new VIN system, still using thirteen digits but, most important, including a number code identifying the engine the car was equipped with. Until 1971, you could not tell which engine came in a particular car, only whether a car was equipped with a four-, six- or eight-cylinder engine. Chevrolet stamped the last eight digits of a car's VIN on the engine, thereby "matching" the engine to the car. Unfortunately, it is easy to restamp a cylinder block with the VIN, thereby increasing a car's value.

A car's VIN was stamped on a metal plate and attached to the driver's-side door pillar until 1967. From 1968 on, the number was stamped on a plate and attached to the left side of the dash, making it visible through the windshield.

Each chapter of this book shows a complete VIN breakdown and, by doing so, indicates on which models the SS Option Package was available when the Super Sport was not a separate model series.

Besides original pricing, quantities of the options sold are included. These have been compiled in a 2,000 page book published by Tailfins and Bowties (P.O. Box 249, Glenpool, Oklahoma 74033) from Chevrolet records and used by the publisher's permission.

Chevrolet kept only cumulative totals for the number of options installed on all models within a car line. Unfortunately, it did not go a step further and indicate how many of a particular combination of options were made—for example, how many Super Sport Camaros also include the Rally Sport option. It has been estimated that the number of possible combinations resulting from Chevrolet models, colors, trims and options—with no two alike—is 5,153,859,497,004,179,760.

Another point to keep in mind is that the total shown for each option may not be the true total. For example, the F41 Suspension Package was standard with the Super Sport Chevelle in 1970. The total under the F41 option does not include the F41 Packages installed with the Super Sport Chevelles, but rather the packages installed on all other Chevelles sold that year.

Even if all the numbers match on a particular car you are looking at, especially on one built before 1972, it is to your advantage if the car is documented. It is all the better if the previous owner can provide you with the original invoice or window sticker, any service records, a Protect-O-Plate and the like. This is especially important with the rare, popular models, such as the LS6 Chevelles.

The colors and interior trim listed in each chapter are correct as far as they go. However, Chevrolet did build cars in colors and trim combinations not listed. As with all the information listed here, be open to the possibility that exceptions can and do occur. This means that you'll have to work harder to determine authenticity.

Every effort has been made to make sure that the information contained in this book is correct. However, I would like to hear from enthusiasts with any corrections, variations, or interesting additions. Please write to me care of Motorbooks International.

Chapter 1

1961 Impala SS

Production
SS Option Package,
all body styles   456

Serial numbers

Description
11869F100001
1 — Last digit of model year
1869 — Model number (1811-2 dr sedan, 1837-2 dr coupe,
   1839-4 dr hardtop, 1867-2 dr convertible, 1869-4 dr sedan)
F — Assembly plant (A-Atlanta, B-Baltimore, F-Flint, J-Janeville,
   K-Kansas City, L-Los Angeles, N-Norwood, O-Oakland,
   S-St Louis, T-Tarrytown, W-Willow run)
100001 — Consecutive sequence number

Location
   On plate attached to left front door hinge post.

Engine and transmission suffix codes
FH — 348 ci 3x2V V-8 350 hp, 4 speed manual
FJ — 348 ci 4V V-8 340 hp, 4 speed manual
FL — 348 ci 4V V-8 305 hp, 4 speed manual
GE — 348 ci 4V V-8 305 hp, Powerglide automatic
Q, QA — 409 ci 4V V-8 360 hp, 4 speed manual

Carburetors
348 ci 340 hp — 3772600
348 ci 350 hp — 7013015 front, 7013020 center,
   7013026 center w/Turboglide, 7013015 rear
409 ci — 3794438

Options*

1811 2 dr sedan, 8 cyl	$2,643.00
1837 2 dr sport coupe, 8 cyl	2,704.00
1839 4 dr sport sedan, 8 cyl	2,769.00
1867 2 dr convertible, 8 cyl	2,954.00
1869 4 dr sedan, 8 cyl	2,697.00

Option number	Description	Retail price
101	Deluxe outside air heater & defroster	$ 74.25
103	Manual control radio	53.80
104	Push-button control radio	62.45
110	Deluxe AC	457.30

111	Cool-Pack AC	317.45
116	Recirculating inside air heater & defroster	46.85
117	4 wheel covers (w/7.50–14 or 8.00–14 tires only)	14.55
121	Temperature-controlled radiator fan	16.15
139	Rear compartment lock	
	6 passenger station wagons	10.80
	9 passenger station wagons	16.15
140	Body Equipment Group A	48.45
145	Body Equipment Group B	
	2 dr models	23.15
	4 dr models	25.85
146	Speed & cruise control	94.70
200	HD front & rear shock absorbers (exc station wagons)	1.10
220	Dual exhaust (for Economy Turbo-Fire or Super Turbo-Fire engines only)	24.75
242	Special crankcase ventilation (for state of Calif only)	5.40
253	HD Coil front springs	1.10
257	HD radiator	10.80
259	Divided second seat (for station wagons only)	37.70
302	Turboglide transmission	209.85
313	Powerglide transmission	199.10
315	Overdrive transmission	107.60
324	Power steering	75.35
326	40 amp generator (NA w/305 hp, 340 hp, 350 hp engines)	26.90
331	Tachometer (mounted on steering column)	48.45
333	2 speed electric windshield wipers (incl windshield washers)	17.25
335	Foam rubber front seat cushion	7.55
338	35 amp generator (NA w/305 hp, 340 hp, 350 hp engines)	7.55
345	HD 12 volt 66 plate 70 amp-hr battery	7.55
347	Deluxe equipment	16.15
348	Deluxe steering wheel (Biscayne & Brookwood only)	3.80
378	50 amp, low-cut-in generator (NA w/305 hp, 340 hp, 350 hp engines; NA w/AC)	96.85
380	Power seat	96.85
398	E-Z Eye tinted glass	
	All windows (all models)	37.70
	Windshield (all models)	21.55
	Rear window only (models 1637–1837)	14.00
410	230 hp Super Turbo-Fire engine (incl 4 bbl carburetor)	29.10

412	Vacuum power brake	43.05
424	Electric control power rear window (for 6 passenger station wagons only; std on 9 passenger models)	32.30
426	Electric control power windows (exc Biscayne, Biscayne Fleetmaster & Brookwood models)	102.25
427	Padded instrument panel	18.30
572	305 hp Special Turbo-Thrust engine (incl 4 bbl carburetor, dual exhaust & temperature-controlled radiator fan)	209.85
573 & 576	280 hp Super Turbo-Thrust engine (incl 3 2 bbl carburetors, dual exhaust, temperature-controlled radiator fan)	163.60
590	340 hp Special Turbo-Thrust engine (incl 4 bbl carburetor, dual exhaust, mechanical valve lifters; temperature controlled radiator fan & high-lift camshaft)	236.75
590 & 593	350 hp Special Super Turbo-Thrust engine (incl 3 2 bbl carburetors, dual exhaust, high-lift camshaft, temperature controlled radiator fan & mechanical valve lifters)	258.30
593	HD coil rear springs	2.70
675	Positraction limited-slip rear axle	43.05
685	4 speed close-ratio synchromesh transmission	188.30
686	HD brakes w/metallic facings (NA w/taxi equipment or police car chassis)	37.70
808	Gray vinyl trim	5.40
	Exterior paint	
	Solid colors	NC
	Two-tone combinations	16.15

*Options are for Biscayne, Bel Air and Impala.

Facts

A midyear option on the 1961 Impala was the Super Sport. Available on all Impala body styles, it consisted of upgraded tires on station wagon wheels, springs, shocks and special sintered metallic brake linings. Power steering and power brakes were also included. Engine choice was limited to the four larger-output V-8s: three single four-barrel 348 ci engines, two rated at 305 hp and one rated 340 hp, and a 3x2 barrel version rated at 350 hp.

The highest-output engine available on the Impala was the new for 1961 409 ci V-8. A bored and stroked version of the 348, it was rated at 360 hp on a single four-barrel carburetor and was known as the Turbo-Fire 409. A true performance engine, the 409 came with an aluminum intake manifold, a Carter four-barrel carburetor, a solid-lifter camshaft and an 11.25:1 compression ratio. A total of 142 cars got this engine in 1961.

In the interior, the SS Package consisted of a passenger's-side grab bar with Impala SS script, a 7000 rpm steering-wheel-mounted tachometer, a shifter plate for four-speed-manual-equipped Impalas and a dash panel pad. The Impala SS could be identified by SS emblems on the rear fenders and trunk lid. Wheel covers were unique, featuring three-blade knock-off-type spinners.

A total of 456 Super Sport equipped Impalas were produced.

The 1961 Impala SS Convertible.

Chapter 2

1962 Impala SS

Production
SS equipment option   99,311

Serial numbers

Description
21747F100001
2 — Last digit of model year (1962)
1747 — Model number (1747-2 dr sedan, 1767-2 dr convertible,
    1847-2 dr sedan, 1867-2 dr convertible)
F — Assembly plant (A-Atlanta, B-Baltimore, F-Flint, J-Janeville,
    K-Kansas City, L-Los Angeles, N-Norwood, O-Oakland,
    S-St Louis, T-Tarrytown, W-Willow Run)
100001 — Consecutive sequence number

Location
On plate attached to left front door hinge post.

Engine and transmission suffix codes
A, AE, AK, AM — 235 ci I-6 1 bbl 135 hp, 3 speed manual
AF, AG, AJ, AZ — 235 ci I-6 1 bbl 135 hp, 3 speed manual w/AC
BG, BH — 235 ci I-6 1 bbl 135 hp, Powerglide automatic w/AC
C — 283 ci V-8 2 bbl 170 hp, 3 speed manual
CD — 283 ci V-8 2 bbl 170 hp, overdrive
CL — 283 ci V-8 2 bbl 170 hp, 3 speed manual w/AC
D — 283 ci V-8 2 bbl 170 hp, Powerglide automatic
DK — 283 ci V-8 2 bbl 170 hp, Powerglide automatic w/AC
QA — 409 ci V-8 4 bbl 380 hp, 4 speed manual
QB — 409 ci V-8 2x4 bbl 409 hp, 4 speed manual
R — 327 ci V-8 4 bbl 250 hp, 3 or 4 speed manual
RA — 327 ci V-8 4 bbl 250 hp, 3 or 4 speed manual w/AC
RB — 327 ci V-8 4 bbl 300 hp, 3 or 4 speed manual
S — 327 ci V-8 4 bbl 250 hp, Powerglide automatic
SA — 327 ci V-8 4 bbl 250 hp, Powerglide automatic w/AC
SB — 327 ci V-8 4 bbl 300 hp, Powerglide automatic

Carburetors
327 ci 4 bbl — 7020006, 7020012
409 ci 4 bbl — 3820580
409 ci 2x4 bbl —3815403 front, 3856580 rear

Exterior color codes

Tuxedo Black	900	Silver Blue	912
Surf Green	903	Nassau Blue	914
Laurel Green	905	Twilight Turquoise	917

Exterior color codes

Twilight Blue	918	Ermine White	936
Autumn Gold	920	Adobe Beige	938
Roman Red	923	Satin Silver	940
Coronna Cream	925	Honduras Maroon	948
Anniversary Gold	927		

Two-tone color codes

Ermine White/ Tuxedo Black	950
Ermine White/Surf Green	953
Surf Green/Laurel Green	955
Ermine White/Silver Blue	959
Silver Blue/Nassau Blue	962
Ermine White/ Twilight Blue	963
Twilight Turquoise/ Twilight Blue	965
Adobe Beige/ Autumn Gold	970
Ermine White/Roman Red	973
Ermine White/Satin Silver	984

Interior trim codes

Color	Coupe	Convertible
Black	—	814/815*
Green	826/827*	829/821*
Blue	842/843*	836/831*
Aqua	853/854*	847/845*
Fawn	866/867*	870/856*
Red	847/875*	886/879*
Gold	891/892*	894/890*

*Bucket seats mandatory with Super Sport.

Convertible top color codes

White	Std
Black	470A
Cream	470B
Blue	470C

Options*

1747 2 dr sport coupe, 6 cyl	$2,669.00
1767 2 dr convertible, 6 cyl	2,919.00
1847 2 dr sport coupe, 8 cyl	2,776.00
1867 2 dr convertible, 8 cyl	3,026.00

Option number	Description	Retail price
103	Manual control radio	$ 47.90
104	Push-button control radio	56.50
110	Deluxe AC (incl 42 amp Delcotron & HD radiator)	363.70
111	Cool-Pack AC (incl HD radiator)	317.45
117	4 wheel covers (w/7.00-14, 7.50-14 or 8.00-14 tires only)	18.30
121	Temperature-controlled radiator fan	16.15
139	Rear compartment lock (station wagons only)	10.80
147	Comfort & convenience equipment	
	Impala Series	30.15
	Bel Air Series	40.90
	Biscayne Series	44.15
148	Front seatbelts	
	Driver only	10.25
	Driver & passenger	18.90

149	Front grille guard	15.10
150	Rear bumper guard (exc station wagons)	9.70
200	HD front & rear shock absorbers (std on station wagons)	1.10
203	3.08 ratio rear axle (for use w/4 speed transmission & 380 hp or 409 hp engines only)	NC
216	Oil bath air cleaner (1 pt capacity)	5.40
227	HD clutch (incl w/taxi equipment)	5.40
242	Special crankcase ventilation (for state of Calif only)	5.40
253	HD coil front springs	1.10
257	HD radiator	10.80
259	Divided second seat (for station wagons only)	37.70
300	250 hp Turbo-Fire 327 engine (incl 4 bbl carburetor, dual exhaust, temperature-controlled radiator fan)	83.95
313	Powerglide transmission	188.30
315	Overdrive transmission	107.60
317	42 amp Delcotron generator	26.90
324	Power steering	75.35
331	Tachometer (mounted on steering column)	48.45
333	2 speed electric windshield wipers (incl washers)	17.25
335	Foam rubber front seat cushion (Biscayne only)	7.55
338	35 amp generator	7.55
345	HD 12 volt 66 plate 70 amp-hr battery	7.55
348	Deluxe steering wheel (Biscayne only)	3.80
380	6 way electric control power seat (exc Biscayne models; front seat only; NA w/SS equipment)	96.85
397	300 hp Turbo-Fire 327 engine (incl large 4 bbl carburetor, dual exhaust, temperature-controlled radiator fan)	137.75
398	Soft Ray tinted glass	
	All windows (all models)	37.70
	Windshield only (all models)	21.55
	Rear window only (model 1537)	14.00
412	Vacuum power brakes	43.05
417	Positive engine ventilation	11.85
424	Electric control power rear window (for 6 passenger station wagons only; std on 9 passenger models)	32.30
426	Electric control power windows (exc Biscayne models)	102.25
427	Padded instrument panel	16.15
434	52 amp Delcotron generator (NA w/taxi equipment)	

	For use wo/deluxe AC	34.45
	For use w/deluxe AC	7.55
435	62 amp Delcotron generator	
	For use wo/deluxe AC	96.85
	For use w/deluxe AC	69.95
580	380 hp Turbo-Fire 409 engine (incl large 4 bbl carburetor, dual exhaust, mechanical valve lifters, temperature-controlled radiator fan, high-lift camshaft, 8.00–14 4 ply regular highway blackwall tires, HD front & rear springs & shock absorbers)	320.65
580–	409 hp Turbo-Fire 409 engine (incl dual	
587	4 bbl carburetors, dual exhaust, mechanical valve lifters, temperature-controlled radiator fan, high-lift camshaft, 8.00–14 4 ply regular highway blackwall tires, HD front & rear springs & shock absorbers)	376.65
581	Economy carburetor (Biscayne only; not recommended for highway speeds)	7.55
593	HD coil rear springs (std on 9 passenger station wagons)	2.70
675	Positraction rear axle	43.05
685	4 speed close-ratio synchromesh transmission	
	w/250 hp engine	188.30
	w/300 hp, 380 hp or 409 hp engine; incl tachometer	236.75
686	HD brakes w/metallic facings	37.70
865	Vinyl trim	11.85
	Exterior paint	
	single colors	NC
	Two-tone combinations	16.15

*Options are for Biscayne, Bel Air and Impala.

Facts

The use of the Super Sport option expanded in 1962, but only on two body styles, the two-door coupe and convertible. Unlike the SS Option Package of 1961, it could be had with any engine, from the standard 235 ci six-cylinder to the 409 ci big-block V-8. The 348 ci V-8s were dropped, replaced by 327 ci small-blocks. Horsepower ratings on the two optional 409 ci V-8s were upped to 380 on the single four-barrel and 409 on the 2x4 barrel.

The V-8 powered Impalas were a separate model series from six-cylinder models, but the numbers used to designate body style and model series were the same.

The heavy-duty mechanical items of the Super Sport option (heavy-duty springs, shocks, brake linings and so on) were deleted in 1962, though these items were available optionally. In this guise, the Super Sport option was just a trim package.

In the interior, the passenger assist bar was still used, as was the chrome shift plate on cars equipped with floor-shifted manual transmissions. Vinyl bucket seats were standard equipment, as was a center locking console. The 7000 rpm tachometer, standard in 1961, was not relegated to the option list.

On the exterior, the most noticeable difference between the Super Sport Impala and the regular Impalas was the use of aluminum body molding inserts; plain Impalas came with painted inserts. The rear fender SS emblems were redesigned. For 1962, SS letters with red inlay were positioned over the circular Impala emblem. An Impala SS emblem was located on the right rear part of the trunk lid. Wheel covers were the regular Impala units with the three-blade spinner.

The 1962 Impala SS Convertible at the Chicago Auto Show along with the Corvair and Chevy II Nova.

Chapter 3

1963 Impala SS

Production
SS equipment option   153,271

Serial numbers

Description
31747F100001

3 — Last digit of model year (1963)

1747 — Model number (1747-2 dr sedan, 1767-2 dr convertible, 1847-2 dr sedan, 1867-2 dr convertible)

F — Assembly plant (A-Atlanta, B-Baltimore, F-Flint, J-Janeville, K-Kansas City, L-Los Angeles, N-Norwood, O-Oakland S-St Louis, T-Tarrytown, W-Willow Run)

100001 — Consecutive sequence number

Location
On plate attached to left front door hinge post.

Engine and transmission suffix codes
A, AE — 230 ci I-6 1 bbl 140 hp, 3 speed manual

B — 230 ci I-6 1 bbl 140 hp, Powerglide automatic

C — 283 ci V-8 2 bbl 195 hp, 3 speed manual

CB — 283 ci V-8 2 bbl 195 hp, police

CD — 283 ci V-8 2 bbl 195 hp, overdrive

D — 283 ci V-8 2 bbl 195 hp, Powerglide automatic

DK — 283 ci V-8 2 bbl 195 hp, Powerglide automatic w/AC

QA — 409 ci V-8 4 bbl 400 hp, 4 speed manual

QB — 409 ci V-8 2x4 bbl 425 hp, 4 speed manual

QC — 409 ci V-8 4 bbl 340 hp, 4 speed manual

QG — 409 ci V-8 4 bbl 340 hp, Powerglide automatic

QM — 427 ci V-8 2x4 bbl 430 hp, 4 speed manual

R — 327 ci V-8 4 bbl 250 hp, 3 or 4 speed manual

RA — 327 ci V-8 4 bbl 250 hp, 3 or 4 speed manual w/AC

RB — 327 ci V-8 4 bbl 300 hp, 3 or 4 speed manual

S — 327 ci V-8 4 bbl 250 hp, Powerglide automatic

SA — 327 ci V-8 4 bbl 250 hp, Powerglide automatic w/AC

SB — 327 ci V-8 4 bbl 300 hp, Powerglide automatic

SG — 327 ci V-8 4 bbl 300 hp, Powerglide automatic w/AC

XE — 327 ci V-8 4 bbl 250 hp, service replacement block

Carburetors
327 ci 250 hp — 3846247

327 ci 250 hp w/Powerglide — 7013006, 7013012, 3846246

327 ci 300 hp — 3851761

327 ci 300 hp w/Powerglide — 3851762

409 ci 400 hp — 3855581
409/427 ci 425/430 hp — 3815403 front, 3856580 rear

Exterior color codes

Tuxedo Black	900
Laurel Green	905
Ivy Green	908
Silver Blue	912
Monaco Blue	914
Azure Aqua	918
Marine Aqua	919
Autumn Gold	920
Ember Red	922
Saddle Tan	932
Cordovan Brown	934
Ermine White	936
Adobe Beige	938
Satin Silver	940
Palomar Red	948

Two-tone color codes

Ermine White/ Tuxedo Black	950
Ermine White/ Laurel Green	954
Ermine White/Silver Blue	959
Silver Blue/Monaco Blue	962
Ermine White/Azure Aqua	963
Azure Aqua/Marine Aqua	967
Adobe Beige/Autumn Gold	970
Adobe Beige/Saddle Tan	971
Adobe Beige/ Cordovan Brown	972
Ermine White/Ember Red	973
Ermine White/Satin Silver	984

Interior trim codes

Color	Coupe	Convertible
Black	814/815*	—
Green	829/821*	826
Blue	836/831*	842
Aqua	847/845*	853
Saddle	859/852*	857
Fawn	870/856*	866
Red	886/879*	874

*Bucket seats optional with Super Sport.

Convertible top color codes

White	Std
Black	CO5-A
Beige	CO5-B

Options*

1747 2 dr sport coupe	$2,667.00
1767 2 dr convertible	2,917.00
1847 2 dr sport coupe	2,774.00
1867 2 dr convertible	3,024.00

Option number	Description	Quantity	Retail price
A01	Soft Ray tinted glass (all windows)	254,973	$ 37.70
A02	Soft Ray tinted glass (windshield only)	484,438	21.55
A31	Power windows (NA on Biscayne models)	26,857	102.25
A33	Power rear window (station wagons only)	48,570	32.30
A37	Front seatbelts	254,523	
	Driver only		10.25
	Driver & passenger		18.85

A42	6 way electric control power seat (front seat only; NA w/SS equipment)	14,016	96.85
A66	Divided second seat (for station wagons; Fawn trim only)	1,637	37.70
A96	Stowage compartment lock (station wagons only)	11,215	10.80
B50	Foam rubber front seat cushion (for Biscayne only)	34,953	7.55
B70	Padded instrument panel	536,234	18.30
C08	Vinyl roof cover (for model 1747 only; solid exterior colors only)	10,468	
	Black vinyl		75.35
	White vinyl		75.35
C14	2 speed electric windshield wipers & washer (incl when comfort & convenience equipment is ordered)	773,806	17.25
C48	Heater & defroster deletion (NA w/AC)	21,651	72.00 CR
C60	Four Season AC (incl 52 amp Delcotron, HD radiator & temperature-controlled radiator fan; 7.50-14 or larger tires required)	194,825	363.70
C65	Custom AC (incl 52 amp Delcotron & HD radiator; 7.50-14 or larger tires required)	—	317.45
F40	Special suspension (NA w/SS equipment; incl w/taxi & police car chassis equipment; on sedans, sport coupe & convertible, incl HD front & rear springs & HD front & rear shock absorbers)	—	4.85
F60	HD front springs (on 9 passenger station wagons only)	18,522	1.10
G76	3.36 ratio rear axle	2,860	2.20
G80	Positraction rear axle (NA w/taxi or police car chassis)	98,022	
	3.08 ratio		43.05

	3.36 ratio		43.05
	3.55 ratio		43.05
	3.70 ratio		43.05
	4.11 ratio		43.05
	4.56 ratio		43.05
G96	3.55 ratio rear axle	10,714	NC
J50	Vacuum power brakes	264,392	43.05
J65	Special brakes w/metallic facings (NA w/taxi equipment; incl on police car chassis)	3,511	37.70
K02	Temperature-controlled radiator fan (incl when Four Season AC is ordered)	2,501	16.15
K23	Positive engine ventilation	1,163	6.50
K45	Oil bath air cleaner (1 pt capacity)	35,892	5.40
K79	42 amp Delcotron alternating current generator	4,371	10.80
K81	62 amp Delcotron alternating current generator	2,727	
	For use wo/AC		75.35
	For use w/AC		64.60
K82	52 amp Delcotron alternating current generator (NA w/taxi equipment; incl w/AC)	2,737	21.55
M01	HD clutch (incl when taxi equipment is ordered)	4,398	5.40
M10	Overdrive transmission	16,558	107.60
M20	4 speed synchromesh transmission w/std-ratio 2.54:1 low gear	39,063	
	When 250 hp engine is ordered		188.30
	When 300 hp, 340 hp 400 hp or 425 hp engine is ordered; incl tachometer		236.75
M21	4 speed synchromesh transmission w/close-ratio 2.20:1 low gear (w/400 hp & 425 hp engines only; incl tachometer)	4,075	236.75
M35	Powerglide transmission	1,169,786	188.30
N30	Bel Air type deluxe steering wheel w/half-horn ring (for Biscayne only)	103,234	3.80

Code	Description		
N33	7 position Comfortilt steering wheel (available only on Bel Air & Impala series when ordered w/power steering & Powerglide transmission)	17,817	43.05
N40	Power steering	744,559	75.35
P01	4 brightmetal wheel covers (for use only w/14 in. wheels; NA w/SS equipment)	615,707	18.30
P02	Simulated wire wheel covers	40,047	
	Used w/SS equipment		24.75
	Used wo/SS equipment		43.05
P05	Set of 5 14x5 chrome wheels (NA on station wagons)	893	
	w/SS equipment		46.30
	wo/SS equipment		64.60
P12	Set of 5 14x6JK wheels (std on station wagons; 8.00–14 tires required)	1,094	5.40
T60	HD 66 plate 70 amp-hr battery	32,259	7.55
U16	Tachometer (mounted on steering column; incl when 4 speed transmission is ordered w/300 hp, 340 hp, 400 hp & 425 hp engines)	2,357	48.45
U60	Manual control radio	177,781	47.90
U63	Push-button control radio	484,915	56.50
U69	Push-button control AM/FM radio	856	134.50
V01	HD radiator	52,029	10.80
V20	Front grille guard	88,033	19.40
V32	Rear bumper guard (NA on station wagons)	69,285	9.70
V55	Luggage carrier (wagons only)	21,626	43.05
Z01	Comfort & convenience equipment	294,470	
	Impala Series		30.15
	Bel Air Series		40.90
	Biscayne Series		44.15
Z02	Push-button control radio & rear speaker	119,924	69.95

Z03	SS equipment (incl all-vinyl interior trim, front bucket seats, special emblems, wheel covers [14 in. wheels only], side molding inserts & special front seats; w/std 3 speed or overdrive transmission)	153,271	161.40
Z04	Police car chassis equipment	447	—
Z05	Economy carburetor (Biscayne only)	439	5.40
Z10	Push-button control AM/FM radio w/rear seat speaker	—	147.95
865	Vinyl trim	—	5.40
	Exterior paint		
	Solid colors		NC
	Two-tone combinations		16.15

*Options are for Biscayne, Bel Air and Impala.

Facts

The 1963 Chevrolet line was restyled. Although the car's look was similar, its cleaner lines were supposed to make it appear more luxurious. The Super Sport option was again a trim package. On the exterior, Super Sport Impalas got the anodized aluminum side and rear trim panel inserts. The SS identification differed from that in 1962: the red-filled rear quarter SS letters were located above the Impala script and no lettering was used on the truck lid. Wheel covers again got the nonfunctional spinners.

In the interior, all Super Sport Impalas came with vinyl bucket seats separated by a new console that had SS identification in front of the shifter. The dash panel was all-new and Super Sport models got anodized panel inserts. The SS identification was included on the steering wheel.

Option number codes were changed starting in 1963. Before 1963, an option could be identified by three numbers; from 1963, options were identified by a letter followed by two numbers.

Engine availability was essentially the same as in 1962; however, a low-horsepower version of the 409 was added, rated at 340 hp. Unlike the more powerful 409s, it came with a hydraulic camshaft. The more powerful 409s were now rated at 400 hp and 425 hp, and were available only with a four-speed manual transmission. A total 16,802 of the 409 engines were installed, most of these in Super Sport optioned Impalas.

The high point in 1963 was reached with the introduction of the Z11 option. At $1,245 on the sport coupe model, it was intended solely for drag racing. The Impala's weight was reduced through the use of an aluminum front end, saving 112 lb, and the deletion of other front bracketry and bracing further reduced the car's weight by 121 lb. The Z11 used a bored and stroked version of the 409 engine, now displacing 427 ci. Rated at 430 hp, the 427 got a new isolated-runner dual-quad aluminum intake manifold. A total of fifty-seven of the Z11 equipped Impalas were built.

The 409 ci based 427 ci engine should not be confused with the 396 ci based 427 ci engine issued in 1966. Although they both have the same bore and stroke, the later 427s were based on the Mark IV engine family, which used highly efficient canted valve cylinder heads.

The 1963 Impala SS Convertible.

Chapter 4

1964 Impala SS

Production

1347 2 dr coupe, 6 cyl	1,998	
1367 2 dr convertible, 6 cyl	316	
1447 2 dr coupe, 8 cyl	97,753	
1467 2 dr convertible, 8 cyl	19,099	
Total	119,166	

Serial numbers

Description

41347F100001

4 — Last digit of model year (1964)

1347 — Model number (1347–2 dr coupe, 1367–2 dr convertible, 1447–2 dr coupe, 1467–2 dr convertible)

F — Assembly plant (A-Atlanta, B-Baltimore, F-Flint, J-Janeville, K-Kansas City, L-Los Angeles, N-Norwood, O-Oakland, S-St Louis, T-Tarrytown, W-Willow Run)

100001 — Consecutive sequence number

Location

On plate attached to left front door hinge post.

Engine and transmission suffix codes

A, AE — 230 ci I-6 1 bbl 140 hp, 3 speed manual

AF, AG — 230 ci I-6 1 bbl 140 hp, 3 speed manual w/AC

B — 230 ci I-6 1 bbl 140 hp, Powerglide automatic

BQ — 230 ci I-6 1 bbl 140 hp, Powerglide automatic w/AC

C — 283 ci V-8 2 bbl 195 hp, 3 speed manual

CB — 283 ci V-8 2 bbl 195 hp, police

D — 283 ci V-8 2 bbl 195 hp, Powerglide automatic

QA — 409 ci V-8 4 bbl 400 hp, 4 speed manual

QB — 409 ci V-8 2x4 bbl 425 hp, 4 speed manual

QC — 409 ci V-8 4 bbl 340 hp, 3 or 4 speed manual

QN — 409 ci V-8 4 bbl 400 hp, 4 speed manual w/transistor ignition

QP — 409 ci V-8 2x4 bbl 425 hp, 4 speed manual w/transistor ignition

QQ — 409 ci V-8 4 bbl 340 hp, 3 or 4 speed manual w/transistor ignition

QR — 409 ci V-8 4 bbl 340 hp, Powerglide automatic w/transistor ignition

R — 327 ci V-8 4 bbl 250 hp, 3 or 4 speed manual

RB — 327 ci V-8 4 bbl 300 hp, 3 or 4 speed manual

S — 327 ci V-8 4 bbl 250 hp, Powerglide automatic

SB — 327 ci V-8 4 bbl 300 hp, Powerglide automatic

XE — 327 ci V-8 4 bbl 300 hp, service replacement block

Carburetors

327 ci 250 hp — 3846247
327 ci 250 hp w/Powerglide — 3846246
327 ci 300 hp — 3851761
327 ci 300 hp w/Powerglide — 3851762
409 ci 400 hp — 3855581
409 ci 425 hp — 3815403 front, 3856580 rear

Distributors

327 ci — 1111016
409 ci 340 hp — 1111023
409 ci 340 hp w/transistor
  ignition — 1111059
409 ci 400/425 hp w/transistor
  ignition — 1111086

Exterior color codes

Tuxedo Black	900
Meadow Green	905
Bahama Green	908
Silver Blue	912
Daytona Blue	916
Azure Aqua	918
Lagoon Aqua	919
Almond Fawn	920
Ember Red	922
Saddle Tan	932
Ermine White	936
Desert Beige	938
Satin Silver	940
Goldwood Yellow	943
Palomar Red	948

Two-tone color codes

Bahama Green/ Meadow Green	952
Ermine White/ Meadow Green	954
Ermine White/ Silver Blue	959
Daytona Blue/ Silver Blue	960
Ermine White/ Lagoon Aqua	965
Desert Beige/Saddle Tan	971
Desert Beige/Ember Red	975
Daytona Blue/ Satin Silver	982
Azure Aqua/Ermine White	988
Desert Beige/Palomar Red	993
Satin Silver/Palomar Red	995

Interior trim codes

Silver/Black	805
Black	815
White/Aqua	845
Fawn	856
Saddle	862
White/Red	878
Red	879

Convertible top color codes

White	Std
Beige	AB
Black	AA

Options

1347 2 dr sport coupe	$2,839.00
1367 2 dr convertible	3,088.00
1447 2 dr sport coupe	2,947.00
1467 2 dr convertible	3,196.00

Option number	Description	Quantity	Retail price
A01	Soft Ray tinted glass (all windows)	278,038	$ 37.70

Option number	Description	Quantity	Retail price
A02	Soft Ray tinted glass (windshield only)	552,520	21.55
A31	Power windows (NA on Biscayne Series)	31,410	102.25
A33	Power rear window (wagons only)	51,268	32.30
A37	Custom deluxe front seatbelts (driver & passenger)	240,448	3.25
A42	6 way electric control power seat (front seat only; NA on Impala SS or Biscayne Series)	14,257	96.85
A49	Custom deluxe front seatbelts w/retractors (driver & passenger)	257,564	7.55
A62	Front seatbelts deletion (driver & passenger)	—	11.00 CR
A66	Divided second seat (for station wagons; Fawn trim only)	1,157	37.70
A96	Stowage compartment lock (station wagons only)	11,540	10.80
B01	HD body equipment (for Biscayne only)	350	10.80–18.30
B02	Taxi equipment	1,743	63.50
B50	Foam rubber front seat cushion (for Biscayne only)	16,785	7.55
B70	Padded instrument panel	622,116	18.30
C05	Convertible tops	40,597	NC
C08	Vinyl roof cover (for models 1347 & 1747 only)	7,326	75.35
C14	2 speed electric windshield wipers & washer (incl when comfort & convenience equipment is ordered)	718,796	17.25
C48	Heater & defroster deletion (NA w/AC)	16,759	72.00 CR
C50	Rear window defroster (sedans & sport coupes only)	—	21.55
C60	Four Season AC	276,772	363.70
C65	Custom deluxe AC	4,209	317.45
F40	Special front & rear suspension (NA on Impala SS Series, taxi or HD chassis equipment)	59,488	

Code	Description	Quantity	Price
	On sedans, sport coupe & convertible, incl special front & rear springs & special front & rear shock absorbers		4.85
	On station wagons, exc 9 passenger, incl special front & rear springs		3.80
F60	Special front springs	2,364	
	9 passenger station wagon		1.10
	6 passenger station wagon when 400 hp or 425 hp engine is ordered		1.10
G66	Superlift rear shock absorbers	5,786	37.70
G76	3.36 ratio rear axle	3,837	2.20
G80	Positraction rear axle	99,834	
	3.08 ratio		43.05
	3.36 ratio		43.05
	3.55 ratio		43.05
	3.70 ratio		43.05
	4.11 ratio		43.05
	4.56 ratio		43.05
G96	3.55 ratio rear axle	10,369	NC
J50	Vacuum power brakes	275,546	43.05
J65	Special brakes w/metallic facings (NA w/taxi equipment)	2,838	37.70
K02	Temperature-controlled radiator fan (incl when Four Season AC is ordered)	2,519	16.15
K23	Closed positive engine ventilation type A	528	6.50
K24	Closed positive engine ventilation type B (approved by state of Calif)	86,737	—
K45	HD air cleaner	24,990	5.40
K66	Full-transistor ignition system (340 hp, 400 hp or 425 hp engine required)	290	75.35
K77	55 amp Delcotron generator (NA w/taxi equipment; incl w/AC)	2,241	21.55
K79	42 amp Delcotron generator	3,220	10.80
K81	62 amp Delcotron generator	2,779	
	For use wo/A/C		75.35

Code	Description	Qty	Price
	For use w/AC		64.60
L30	250 hp Turbo-Fire 327 engine	294,971	94.70
L31	400 hp Turbo-Fire 409 engine	3,044	320.65
L31 or L80	425 hp Turbo-Fire 409 engine	1,997	376.65
L33	340 hp Turbo-Fire 409 engine	5,640	242.10
L74	300 hp Turbo-Fire 327 engine	50,150	137.75
M01	HD clutch (incl when taxi equipment is ordered)	3,461	5.40
M10	Overdrive transmission	12,381	107.60
M20	4 speed synchromesh transmission	46,559	
	When 250 hp engine is ordered		188.30
	When 300 hp, 340 hp, 400 hp or 425 hp engine is ordered; incl tachometer		236.75
M21	4 speed synchromesh transmission (w/400 hp & 425 hp engines only; incl tachometer)	1,945	236.75
M35	Powerglide transmission	1,225,337	199.10
N33	7 position Comfortilt steering wheel (available only on Bel Air, Impala & Impala SS Series when ordered w/power steering & either Powerglide or 4 speed transmission)	40,853	43.05
N34	Sports-styled walnut-grained steering wheel w/plastic rim	5,378	32.30
N40	Power steering	839,498	86.10
P01	4 brightmetal wheel covers (for use only w/14 in. wheels; NA on Impala SS Series)	652,623	18.30
P02	Simulated wire wheel covers (for use only w/14 in. wheels)	29,647	
	Impala SS Series		57.05
	Impala, Bel Air & Biscayne Series		75.35
P12	Set of 5 14x6JK wheels (std on station wagons; 8.00–14 tires required)	5,114	5.40

Code	Description		
T60	HD 66 plate 70 amp-hr battery	62,863	7.55
U16	Tachometer (mounted on instrument panel; incl when 4 speed transmission is ordered w/300 hp, 340 hp, 400 hp & 425 hp engines)	2,407	48.45
U60	Manual control radio	181,026	50.05
U63	Push-button control radio	541,457	58.65
U69	Push-button control AM/FM radio	4,106	136.70
V01	HD radiator (incl when AC is ordered)	45,980	10.80
V31	Front bumper guard	84,930	9.70
V32	Rear bumper guard (NA on station wagons)	71,259	9.70
V55	Luggage carrier (wagons only)	28,950	43.05
Z01	Comfort & convenience equipment type A (incl outside rearview mirror, inside nonglare mirror, 2 speed electric windshield wipers & washer)	399,671	30.15–44.15
Z02	Push-button control radio & rear speaker	141,996	72.10
Z04	HD chassis equipment (Biscayne models only)	620	37.70
Z05	Economy carburetor (Biscayne only)	201	5.40
Z10	Push-button control AM/FM radio w/rear seat speaker	11,896	150.15
Z13	Comfort & convenience equipment type B	39,531	39.85–53.80
865	Vinyl interior trim		5.40
	Exterior paint		
	Single colors		NC
	Two-tone combinations		16.15

*Options are for Biscayne, Bel Air, Impala and Impala SS.

Facts

The Super Sport equipped Impalas became separate models in 1964, rather than an option package.

A slight redesign distinguished the 1964 Impala, giving the car a more boxy, formal look. It was, however, 2 in. narrower and 0.6 in. shorter than the previous year's offerings. The Impala SS got different ent side moldings, which ran along the bodyside sculpturing. The anodized swirl-pattern aluminum inserts began about halfway on the door and continued to the rear fender. The usual red-filled SS letters on the rear fenders did not use the Impala emblem. On the rear, the Super Sport models got the aluminized taillight panel insert and additional SS identification on the right side of the trunk lid. Wheel covers were new. These did not have the nonfunctional knock-off spinners.

The interior, aluminum dash panels set off the Super Sport. The door panels and the top of the redesigned center console carried SS emblems. No SS lettering appeared on the steering wheel.

Engine availability was unchanged from that in 1963. The two larger 409 engines were available only with a four-speed manual transmission. The 340 hp 409 could be had with the two-speed Powerglide transmission. A total 8,684 Impalas were equipped with the 409 engine, and most of these were Super Sport models.

The 1964 Impala SS Convertible.

Chapter 5

1965 Impala SS

Production

16537 2 dr coupe, 6 cyl	3,245	
16567 2 dr convertible, 6 cyl	399	
16637 2 dr coupe,		

8 cyl	212,027
16667 2 dr convertible, 8 cyl	27,443
Total	243,114

Serial numbers

Description

166675F100001

16667 — Model number (16537-2 dr coupe, 16567-2 dr convertible, 16637-2 dr coupe, 16667-2 dr convertible)

5 — Last digit of model year (1965)

F — Assembly plant (A–Atlanta, B–Baltimore, F–Flint, J–Janeville, K–Kansas City, L–Los Angeles, N–Norwood, O–Oakland, S–St Louis, T–Tarrytown, W–Willow Run)

100001 — Consecutive sequence number

Location

On plate attached to left front door hinge post.

Engine and transmission suffix codes

FA, FE — 230 ci I-6 1 bbl 140 hp, 3 speed manual

FF, FL — 230 ci I-6 1 bbl 140 hp, 3 speed manual w/AC

FM — 230 ci I-6 1 bbl 140 hp, Powerglide automatic

FR — 230 ci I-6 1 bbl 140 hp, Powerglide automatic w/AC
250 ci I-6 1 bbl 150 hp

GA — 283 ci V-8 2 bbl 195 hp, 3 speed manual

GC — 283 ci V-8 2 bbl 195 hp, police

GF — 283 ci V-8 2 bbl 195 hp, Powerglide automatic

GK — 283 ci V-8 4 bbl 220 hp, 3 or 4 speed manual

GL — 283 ci V-8 4 bbl 220 hp, Powerglide automatic

HA — 327 ci V-8 4 bbl 250 hp, 3 or 4 speed manual

HB — 327 ci V-8 4 bbl 300 hp, 3 or 4 speed manual

HC — 327 ci V-8 4 bbl 250 hp, Powerglide automatic

HD — 327 ci V-8 4 bbl 300 hp, Powerglide automatic

IA, LF — 396 ci V-8 4 bbl 325 hp, 3 or 4 speed manual

IC — 396 ci V-8 4 bbl 325 hp, 3 or 4 speed manual w/transistor ignition

IE — 396 ci V-8 4 bbl 425 hp, 3 or 4 speed manual

IG, LB — 396 ci V-8 4 bbl 325 hp, Powerglide automatic

II — 396 ci V-8 4 bbl 325 hp, Powerglide automatic w/transistor ignition

IV, LC — 396 ci V-8 4 bbl 325 hp, Turbo Hydra-matic automatic

IW — 396 ci V-8 4 bbl 325 hp, Turbo Hydra-matic automatic w/transistor ignition

JA — 409 ci V-8 4 bbl 400 hp, 3 or 4 speed manual

JB — 409 ci V-8 4 bbl 340 hp, 3 or 4 speed manual

JC — 409 ci V-8 4 bbl 340 hp, 3 or 4 speed manual w/transistor ignition

JD — 409 ci V-8 4 bbl 400 hp, 3 or 4 speed manual w/transistor ignition

JE — 409 ci V-8 4 bbl 340 hp, Powerglide automatic

JF — 409 ci V-8 4 bbl 340 hp, Powerglide automatic w/transistor ignition

Carburetors

396 ci — 3874898 Holley R3139A & 3868864 Holley R3140A

Distributors

327 ci — 1111075
396 ci — 1111073
396 ci w/transistor ignition — 1111098
396 ci 425 hp — 1111100
409 ci — 1111023
409 ci w/transistor ignition — 1111059
409 ci 400 hp — 1111086

Exterior color codes

Tuxedo Black	A
Ermine White	C
Mist Blue	D
Danube Blue	E
Willow Green	H
Cypress Green	J
Artesian Turquoise	K
Tahitian Turquoise	L
Madeira Maroon	N
Evening Orchid	P
Regal Red	R
Sierra Tan	S
Cameo Beige	V
Glacier Gray	W
Crocus Yellow	Y

Convertible top color codes

White	Std
Black	AA
Beige	AB

Options

16537 2 dr coupe	$2,839.00
16567 2 dr convertible	3,104.00

Two-tone color codes*

Ermine White/	
Artesian Turquoise	CK
Mist Blue/Ermine White	DC
Willow Green/	
Ermine White	HC
Cypress Green/Cameo Beige	JV
Tahitian Turquoise/	
Artesian Turquoise	LK
Sierra Tan/Cameo Beige	SV
Cameo Beige/	
Madeira Maroon	VN
Glacier Gray/	
Tuxedo Black	WA
Crocus Yellow/	
Ermine White	YC
*Lower/upper.	

Interior trim codes

Ivory	802
Slate	805
Black	815
Blue	831
Fawn	856
Saddle	862
Red	979

Vinyl top color code

Black	6

Option number	Description	Quantity	Retail price
A01	Soft Ray tinted glass (all windows)	380,092	$ 37.70
A02	Soft Ray tinted glass (windshield only)	586,353	21.55
A31	Power windows	43,700	102.25
A33	Power rear window	56,375	32.30
A42	Power front seat	17,791	96.85
A47	Custom deluxe rear seatbelts	7,378	12.95
A49	Custom deluxe front seatbelts w/retractors	549,662	23.70
A62	Front seatbelts deletion	118,718	11.00 CR
A66	Divided second seat	1,021	37.70
A96	Stowage compartment lock	8,848	10.80
B50	Foam rubber front seat cushion	9,753	7.55
B70	Padded instrument panel	714,220	18.30
C08	Vinyl roof cover	76,347	75.35
C14	Windshield wipers & washer	664,915	17.25
C48	Heater & defroster deletion	12,760	72.00 CR
C50	Rear window defroster	23,343	21.55
C60	Four Season AC	402,565	363.70
F40	Front & rear suspension	84,592	3.80–16.15
F60	Special front springs	2,030	1.10
G66	Superlift rear shock absorbers	20,462	37.70
G67	Superlift rear shock absorbers w/automatic level control	6,850	86.15
G76	3.36 ratio rear axle	2,113	2.20
G80	Positraction rear axle	63,194	43.05
G96	3.55 ratio rear axle	7,437	NC
J50	Vacuum power brakes	370,841	43.05
J65	Special brakes w/metallic facings	2,614	37.70
K02	Temperature-controlled radiator fan	5,080	16.15
K24	Closed positive engine ventilation	133,326	5.40
K45	HD oil bath air cleaner	21,238	5.40
K66	Full-transistor ignition system	589	75.35
K77	55 amp Delcotron generator	2,205	21.55
K79	42 amp Delcotron generator	3,142	10.80
K81	62 amp Delcotron generator	2,501	64.60–75.35

Code	Description	Quantity	Price
L30	327 ci V–8 engine	377,074	92.65
L31	409 ci Hi-Performance V–8 engine	742	320.65
L33	409 ci V–8 engine	2,086	242.10
L35	396 ci V–8 engine	55,454	157.95
L74	327 ci Hi-Performance V–8 engine	56,499	134.80
L78	396 ci V–8 engine	1,838	368.60
M01	HD clutch	3,468	10.80
M10	Overdrive transmission	8,890	107.60
M13	3 speed fully synchronized transmission	1,308	80.70
M20	4 speed synchromesh transmission	60,941	188.30–236.75
M21	4 speed close-ratio synchromesh transmission	1,550	236.75
M35	Powerglide transmission	1,326,468	188.30–199.10
M40	Turbo Hydra-matic transmission	17,821	231.35
M55	Transmission oil cooler	143	16.15
N33	Comfortilt steering wheel	62,057	43.05
N34	Sports-styled walnut-grained steering wheel w/plastic rim	5,864	32.30
N40	Power steering	1,039,252	96.85
P01	4 brightmetal wheel covers	695,172	21.55
P02	4 simulated wire wheel covers	34,336	57.05–75.35
P12	5 14x6JK wheels	8,550	5.40
P19	Spare wheel lock	1,421	5.40
P58	7.35–14–4 pr whitewall tubeless tires	273,112	31.90
P61	7.75–14–4 pr whitewall nylon tubeless tires	10,043	31.80–66.80
P62	7.50–14–4 pr whitewall tubeless tires	586,835	31.80–66.80
P77	8.25–14–4 pr whitewall tubeless tires	214,880	36.15–67.95
T19	8.25–14–8 pr whitewall tubeless tires	1,075	85.05
T60	HD 66 plate 70 amp-hr battery	69,493	7.55
U03	Tri-Volume horn	31,743	14.00
U16	Tachometer	2,957	48.45
U60	Manual control radio	176,325	50.05
U63	Push-button control radio	875,740	58.65
U69	Push-button control AM/FM radio	26,935	136.70
U73	Rear antenna	474,044	NC
U79	Stereo equipment	6,756	107.60

U80	Push-button control AM/FM radio w/rear seat speaker	225,168	150.15
V01	HD radiator	45,168	10.80
V31	Front bumper guard	103,312	16.15
V32	Rear bumper guard	85,971	9.70
Z01	Comfort & convenience equipment type A	611,738	30.15–44.15
Z05	Economy carburetor	161	5.40
Z13	Comfort & convenience equipment type B	48,519	39.85–53.80
Z18	Caprice custom sedan equipment	40,393	242.10
814	Black vinyl interior trim	137,842	10.80
865	Vinyl interior trim	7,745	5.40

Facts

The 1965 Impala SS continued to be a separate model within the Impala Series. The two-door Super Sport coupe could be ordered with the standard six-cylinder engine or with a V–8. The same applied to the Super Sport convertible.

The full-size Chevrolet, along with all other General Motors B-body-based cars, was completely redesigned. It was longer and wider, but the wheelbase remained at 119 in. Giving the two-door coupe a sportier image was the new fastback roof. Not readily visible was the new perimeter frame. The Super Sport Impala could easily be identified by Super Sport script on the front fenders and Impala SS in block letters on the front grille and on the right side of a trim panel running the width of the rear, just above the bumper. This panel was silver on black cars and black on all other cars. New wheel covers had a center-mounted SS emblem. All 1965 Super Sport Impalas came with bright wheelhouse trim.

The interior, too, was all-new. Vinyl bucket seats with bright seatback outline moldings, full carpeting, carpet and vinyl door trim with SS emblems, and a new console that housed a clock and an SS emblem showed off the Super Sport interior. In addition, an aluminum dash panel insert replaced the simulated wood one found on other Impalas.

For the first time, Super Sport models also got full instrumentation as standard equipment, which included oil, amps, temperature and vacuum gauges. The vacuum gauge was replaced by a tachometer on manual-transmission-equipped cars and on V–8 equipped cars with engines rated at 300 hp or more.

As far as engines go, 1965 was a transition year. At model introduction, Super Sport Impalas could be equipped with the same basic line-up as that for 1964 Impalas, with the only deletion being the 425 hp 409 ci V–8. On February 15, 1966, an enlarged version of the 230 ci inline six displacing 250 ci became the standard engine on the six-cylinder Super Sport. More significant was the introduction of the Mark IV big-block, replacing the outdated 409. The 396 ci engine came in two forms. The 325 hp version, designated RPO L735, came with the so-called oval-port cylinder heads, which had 2.06 in. intake and 1.72 in. exhaust valves. Intake manifold was cast in iron and carburetor was a Rochester four-barrel. The L78 designated 396 came with larger rectangular-port cylinder heads, larger 2.19 in. intake valves, a solid-lifter camshaft, an aluminum high-rise intake manifold and a Holley four-barrel carburetor. The new big-block was known as the Turbo-Jet V-8. In addition, a four-barrel version of the 283 ci small-block V-8, rated at 220 hp, was added to the line-up.

The 1965 Impala SS Coupe.

The 1966 Impala SS Convertible.

Chapter 6

1966 Impala SS

Production

16737 2 dr coupe, 6 cyl	823	16867 2 dr convertible,		
16767 2 dr convertible,		8 cyl		15,783
6 cyl	89	Total		119,314
16837 2 dr coupe, 8 cyl	102,619			

Serial numbers

Description

168676F100001

16867 — Model number (16737-2 dr coupe, 16767-2 dr convertible, 16837-2 dr coupe, 16867-2 dr convertible)

6 — Last digit of model year (1966)

F — Assembly plant (A-Atlanta, B-Baltimore, F-Flint, J-Janeville, K-Kansas City, L-Los Angeles, N-Norwood, O-Oakland, S-St Louis, T-Tarrytown, W-Willow Run)

100001 — Consecutive sequence number

Location

On plate attached to left front door hinge post.

Engine and transmission suffix codes

FA, FE — 250 ci I-6 1 bbl 155 hp, 3 speed manual

FL — 250 ci I-6 1 bbl 155 hp, 3 speed manual w/AC

FR — 250 ci I-6 1 bbl 155 hp, Powerglide automatic w/AC

FV — 250 ci I-6 1 bbl 155 hp, 3 speed manual w/AIR

FY — 250 ci I-6 1 bbl 155 hp, 3 speed manual w/AC & AIR

GA — 283 ci V-8 2 bbl 195 hp, 3 speed manual

GC — 283 ci V-8 2 bbl 195 hp, 4 speed manual

GF — 283 ci V-8 2 bbl 195 hp, Powerglide automatic

GQ — 250 ci I-6 1 bbl 155 hp, Powerglide automatic w/AC & AIR

GK — 283 ci V-8 2 bbl 195 hp, 3 speed manual w/AIR

GL — 283 ci V-8 4 bbl 220 hp, Powerglide automatic

GP — 250 ci I-6 1 bbl 155 hp, Powerglide automatic w/AIR

GS — 283 ci V-8 2 bbl 195 hp, 4 speed manual w/AIR

GT — 283 ci V-8 2 bbl 195 hp, Powerglide automatic w/AIR

GW — 283 ci V-8 4 bbl 220 hp, 3 or 4 speed manual

GX — 283 ci V-8 4 bbl 220 hp, 3 or 4 speed manual w/AIR

GZ — 283 ci V-8 4 bbl 220 hp, Powerglide automatic w/AIR

HA — 327 ci V-8 4 bbl 275 hp, 3 or 4 speed manual

HB — 327 ci V-8 4 bbl 275 hp, 3 or 4 speed manual w/AIR

HC — 327 ci V-8 4 bbl 275 hp, Powerglide automatic

HF — 327 ci V-8 4 bbl 275 hp, Powerglide automatic w/AIR

IA — 396 ci V-8 4 bbl 325 hp, 3 or 4 speed manual

IB — 396 ci V-8 4 bbl 325 hp, 3 or 4 speed manual w/AIR

IC — 396 ci V-8 4 bbl 325 hp, Powerglide automatic w/AIR
ID — 427 ci V-8 4 bbl 425 hp, 3 or 4 speed manual
IG — 396 ci V-8 4 bbl 325 hp, Powerglide automatic
IH — 427 ci V-8 4 bbl 390 hp, 3 or 4 speed manual
II — 427 ci V-8 4 bbl 390 hp, Turbo Hydra-matic automatic w/AIR
IJ — 427 ci V-8 4 bbl 390 hp, Turbo Hydra-matic automatic
IN — 396 ci V-8 4 bbl 325 hp, Turbo Hydra-matic automatic w/AIR
IO — 427 ci V-8 4 bbl 425 hp, Turbo Hydra-matic automatic w/AIR
IV — 396 ci V-8 4 bbl 325 hp, Turbo Hydra-matic automatic

Carburetors
396 ci 325 hp — 3874898 (Holley R3139-1AAS)
396 ci 325 hp — 3868854 (Holley R3140-1AAS)
396 ci 325 hp — 3883925 (Holley R3267A) manual
396 ci 325 hp — 3883962 (Holley R3268A) automatic
427 ci 390 hp — 3887147 (Holley R3417A)
427 ci 425 hp — 3885067 (Holley R3246A)

Exterior color codes

Tuxedo Black	A
Ermine White	C
Mist Blue	D
Danube Blue	E
Marina Blue	F
Willow Green	H
Artesian Turquoise	K
Tropic Turquoise	L
Aztec Bronze	M
Madeira Maroon	N
Regal Red	R
Sandalwood Tan	T
Cameo Beige	V
Chateau Slate	W
Lemonwood Yellow	Y

Two-tone color codes*

Ermine White/ Artesian Turquoise	CK
Mist Blue/Ermine White	DC
Mist Blue/Danube Blue	DE
Willow Green/ Ermine White	HC
Tropic Turquoise/ Ermine White	LC
Madeira Maroon/ Tuxedo Black	NA
Sandalwood Tan/ Cameo Beige	TV
Chateau Slate/ Tuxedo Black	WA

Interior trim codes

Black	813
Green	830
Blue	837
Bright Blue	844
Turquoise	846
Fawn	869
Red	873
Ivory	885

Convertible top color codes

White	AA
Black	BB
Beige	—

Vinyl top color codes

Black	BB
Beige	—

Options*

16737 2 dr sport coupe, 6 cyl	$2,822.00/2,842.00**
16767 Convertible, 6 cyl	3,072.00/3,093.00**
16837 2 dr sport coupe, 8 cyl	2,927.00/2,947.00**
16867 Convertible, 8 cyl	3,177.00/3,199.00**

Option number	Description	Quantity	Retail price**
A01	Soft Ray tinted glass (all windows)	389,093	$ 36.60/ 36.90
A02	Soft Ray tinted glass (windshield only)	592,158	20.91/ 21.10
A31	Power windows	43,757	99.33/ 100.10
A33	Power tailgate window	70,093	31.37/ 31.60
A39	Custom deluxe seatbelts w/front retractors	478,960	10.46/ 13.70
A42	6 way power front seat	20,203	94.10/ 94.80
A46	4 way power front seat	3,804	69.01/ 69.55
A51	Strato-Bucket seats	37,421	198.66/ 200.15
A53	Strato-Back seat	14,826	104.56/ 105.35
A66	Divided second seat	839	36.60/ 36.90
A81	Strato-Ease headrests	5,045	52.28/ 52.70
A82	Strato-Ease headrests	2,773	41.82/ 42.15
A96	Stowage compartment lock	7,009	10.46/ 10.55
B01	HD body equipment	367	10.46/ 17.95
B02	Taxi equipment	1,326	56.46/ 62.15
B39	Load floor carpet	5,400	52.28/ 52.70
B50	Foam rubber front seat cushion	6,004	7.32/ 7.40
C08	Vinyl roof cover	199,946	78.42/ 79.00
C48	Heater & defroster deletion	8,786	70.01/ 70.50 CR
C50	Rear window defroster	30,750	20.91/ 21.10
C51	Rear window air deflector	12.441	18.82/ 19.00
C60	Four Season AC	447,011	353.41/ 356.00
C75	Comfortron AC	28,091	416.15/ 419.20
D59	Special instrumentation	14,256	41.82/ 42.15
F40	Special front & rear suspension	84,178	3.66/ 15.80

Code	Description	Quantity	Price
F41	Special purpose front & rear suspension	3,914	31.37/ 31.60
F60	Special front springs	2,573	1.05/ 1.10
G66	Superlift rear shock absorbers	20,661	36.60/ 36.90
G67	Rear shock absorbers w/automatic level control	6,027	83.65/ 84.30
G76	3.36 ratio rear axle	897	2.09/ 2.15
G80	Positraction rear axle	85,877	41.82/ 42.15
H01	3.07 ratio rear axle	91	2.09/ 2.15
H05	3.73 ratio rear axle	223	2.09/ 2.15
J50	Power brakes	363,628	41.82/ 42.15
J65	Metallic brakes	3,385	36.60/ 36.90
K02	Radiator fan	11,169	15.68/ 15.80
K19	AIR equipment	105,475	44.44/ 44.75
K24	Closed positive engine ventilation	6,695	5.23/ 5.25
K30	Speed & cruise control	8,008	75.81/ 76.40
K45	Air cleaner—oil bath	14,902	5.23/ 5.30
K66	Transistorized ignition system	1,530	73.19/ 73.75
K76	61 amp Delcotron generator	3,446	20.91/ 21.10
K79	42 amp Delcotron generator	3,374	10.46/ 10.55
K81	62 amp Delcotron generator	1,948	62.74/ 73.75
L30	275 hp Turbo-Fire 327 engine	366,706	92.01/ 92.70
L35	325 hp Turbo-Jet 396 engine	105,844	156.84/ 158.00
L36	390 hp Turbo-Jet 396 engine	3,287	313.68/ 316.00
L72	425 hp Turbo-Jet 396 engine	1,856	444.38/ 447.65
L77	220 hp Turbo-Fire 283 engine	29,876	52.28/ 52.70
M01	HD clutch	2,983	10.46/ 10.55

Code	Description	Quantity	Price
M10	Overdrive transmission	6,245	115.02/ 115.90
M13	Special 3 speed transmission	1,443	78.42/ 79.00
M20	4 speed transmission	30,467	182.98/ 231.75
M21	4 speed transmission	1,595	230.03/ 231.75
M35	Powerglide transmission	1,193,092	182.98/ 194.85
M40	Turbo Hydra-matic transmission	86,897	224.80/ 226.45
N10	Dual exhaust	58,816	20.91/ 21.10
N33	Comfortilt steering wheel	28,376	41.82/ 42.15
N34	Sports-styled steering wheel	4,183	31.37/ 31.60
N37	Tilt-telescopic steering wheel	17,484	78.42/ 79.00
N40	Power steering	1,039,822	94.10/ 94.80
N96	Mag-style wheel covers	6,642	20.91/ 52.70
P01	Wheel covers	621,646	20.91/ 21.10
P02	Simulated wire wheel covers	21,442	55.42/ 73.75
P12	14x6JK wheels	25,757	5.23/ 5.30
P19	Spare wheel lock	5,450	5.23/ 5.30
T60	HD battery	86,127	7.32/ 7.40
U03	Tri-Volume horn	13,106	13.59/ 13.70
U16	Tachometer	4,075	47.05/ 47.40
U63	Push-button radio	1,026,847	56.99/ 57.40
U69	Push-button AM/FM radio	34,066	132.79/ 133.80
U73	Manual rear antenna	230,428	9.41/ 9.50
U75	Power rear antenna	5,795	28.23/ 28.45
U79	Stereo equipment	12,436	237.35/ 239.15
U80	Auxiliary speaker	227,718	145.86/ 147.00
V01	HD radiator	36,889	10.46/ 10.55

V31	Front bumper guards	83,612	15.68/ 15.80
V32	Rear bumper guards	69,641	9.41/ 9.50
V55	Roof luggage carrier	48,369	41.82/ 42.15
V74	Hazard warning switch equipment	137,218	11.50/ 11.60
Z04	HD chassis equipment	315	36.60/ 36.90
Z19	Comfort & convenience equipment	78,871	18.82/ 26.35
814	Black vinyl interior trim	187,143	10.46/ 10.55
865	Fawn vinyl interior trim	9,258	5.23/ 5.30
950	Two-tone paint combinations	88,594	15.68/ 15.80

*Options are for all Biscayne, Bel Air, Impala and Caprice models.

**Prices reflect a six percent and a seven percent excise tax.

Facts

The 1966 Impala got a slightly revised front grille and a segmented horizontal rear taillight arrangement replacing the six round taillights. The SS identification consisted of the Super Sport script on the front fenders and Super Sport emblems on the grille and on the right side of the trunk lid. The Super Sport Impalas used the same full-size tri-bar Super Sport wheel covers used by the regular Impala.

In the interior, bucket seats were still part of the SS Package, as was the center console. The glovebox door and the console carried SS emblems. The instrument panel gauges were no longer available. Instead, a new optional gauge cluster attached to the console and bottom of the dash panel was available.

Mechanically, all Impalas got a stronger chassis. The optional F41 suspension—which consisted of heavier springs, shocks and sway bars—was a bit ahead of its time, as it included a rear sway bar.

In addition to the 325 hp 396 ci V-8, two 427 ci engines were optionally available. The RPO L36 rated at 390 hp came with the smaller oval-port cylinder heads, hydraulic camshaft and Rochester carburetor. The L72 427 ci came with the larger rectangular-port heads, solid-lifter camshaft, aluminum high-rise intake and Holley 780 cfm carburetor.

Chapter 7

1967 Impala SS and SS 427

Production

16767 2 dr convertible, 6 cyl — 46
16787 & 16887 2 dr coupe, 6 cyl & 8 cyl — 64,387
16867 2 dr convertible, 8 cyl — 9,499
Total — 73,932

Serial numbers

Description

168877F100001

16887 — Model number (16767-2 dr convertible, 16787-2 dr coupe, 16867-2 dr convertible, 16887-2 dr coupe)

7 — Last digit of model year (1967)

F — Assembly plant (C-Southgate, D-Doraville, F-Flint, J-Janeville, L-Los Angeles, R-Arlington, S-St Louis, T-Tarrytown, U-Lordstown, Y-Wilmington, 2-St Therese)

100001 — Consecutive sequence number

Location

On plate attached to left front door hinge post.

Engine and transmission suffix codes

FA, FE — 250 ci I-6 1 bbl 155 hp, 3 speed manual
FF, FL — 250 ci I-6 1 bbl 155 hp, 3 speed manual w/AC
FM — 250 ci I-6 1 bbl 155 hp, Powerglide automatic
FV — 250 ci I-6 1 bbl 155 hp, 3 speed manual w/AIR
FY — 250 ci I-6 1 bbl 155 hp, 3 speed manual w/AC & AIR
GA, GU — 283 ci V-8 2 bbl 195 hp, 3 speed manual
GC — 283 ci V-8 2 bbl 195 hp, 4 speed manual, HD chassis
GF — 283 ci V-8 2 bbl 195 hp, Powerglide automatic
GK — 283 ci V-8 2 bbl 195 hp, 3 speed manual w/AIR
GO, GT — 283 ci V-8 2 bbl 195 hp, Powerglide automatic w/AIR
GP — 250 ci I-6 1 bbl 155 hp, Powerglide automatic w/AIR
GQ — 250 ci I-6 1 bbl 155 hp, Powerglide automatic w/AC & AIR
GS — 283 ci V-8 2 bbl 195 hp, 4 speed manual w/AIR
HA — 327 ci V-8 4 bbl 275 hp, 4 speed manual
HB — 327 ci V-8 4 bbl 275 hp, 4 speed manual w/AIR
HC — 327 ci V-8 4 bbl 275 hp, Powerglide automatic
HF — 327 ci V-8 4 bbl 275 hp, Powerglide automatic w/AIR
IA — 396 ci V-8 4 bbl 325 hp, 3 or 4 speed manual
IB — 396 ci V-8 4 bbl 325 hp, 3 or 4 speed manual w/AIR
IC — 396 ci V-8 4 bbl 325 hp, Powerglide automatic w/AIR
IE, IH — 427 ci V-8 4 bbl 385 hp, 3 or 4 speed manual

IF, IO — 427 ci V-8 4 bbl 385 hp, Turbo Hydra-matic automatic w/AIR
IG — 396 ci V-8 4 bbl 325 hp, Powerglide automatic
II, IX — 427 ci V-8 4 bbl 385 hp, 3 or 4 speed manual w/AIR
IJ, IS — 427 ci V-8 4 bbl 385 hp, Turbo Hydra-matic automatic
IN — 396 ci V-8 4 bbl 325 hp, Turbo Hydra-matic automatic w/AIR
IV — 396 ci V-8 4 bbl 325 hp, Turbo Hydra-matic automatic
KE — 327 ci V-8 4 bbl 275 hp, 4 speed manual
KL — 327 ci V-8 4 bbl 275 hp, Turbo Hydra-matic automatic
KM — 327 ci V-8 4 bbl 275 hp, Turbo Hydra-matic automatic w/AIR

Carburetors
396 ci — 7027201 manual, 7027200 automatic
427 ci — 7027211 manual, 7027210 automatic

Distributors
396 ci — 1111169
427 ci — 1111170

Exterior color codes
Tuxedo Black	AA
Ermine White	CC
Nantucket Blue	DD
Deepwater Blue	EE
Marina Blue	FF
Granada Gold	GG
Mountain Green	HH
Emerald Turquoise	KK
Tahoe Turquoise	LL
Royal Plum	MM
Madeira Maroon	NN
Bolero Red	RR
Sierra Fawn	SS
Capri Cream	TT
Butternut Yellow	YY

Convertible top color codes
White	AA
Black	BB
Medium Blue	—

Two-tone color codes*
Ermine White/ Nantucket Blue	CD
Nantucket Blue/ Ermine White	DC
Nantucket Blue/ Deepwater Blue	DE
Deepwater Blue/ Nantucket Blue	ED
Granada Gold/Capri Cream	GT
Tahoe Turquoise/ Ermine White	LC
Sierra Fawn/Capri Cream	ST

*Lower/upper.

Interior trim codes
Black	810/813
Bright Blue	844/848
Red	870/873
Gold	885/890
Parchment	895/898

Vinyl top color codes
Black	BB
Light Fawn	—

Options*
16767 2 dr convertible, 6 cyl	$3,149.00
16787 2 dr sport coupe, 6 cyl	2,898.00
16867 2 dr convertible, 8 cyl	3,254.00
16887 2 dr sport coupe, 8 cyl	3,003.00

Option number	Description	Quantity	Retail price
A01	Soft Ray tinted glass (all windows)	352,964	$ 36.90

Code	Description	Quantity	Price
A02	Soft Ray tinted glass (windshield only)	469,040	21.10
A31	Power windows	38,473	100.10
A33	Power rear window (wagons)	62,269	31.60
A39	Seatbelts	293,751	6.35–7.90
A42	6 way power front seat	17,983	94.80–105.35
A46	4 way power seat	1,901	69.55
A51	Strato-Bucket seats	12,058	158.00
A53	Strato-Back seat	24,289	115.90
A66	Divided second seat	910	36.90
A75	Low-profile front seat	8,820	11.60–24.80
A76	Rear seat	7,166	6.35
A81	Headrests (w/Strato-Back or Strato-Bucket front seats)	2,752	52.70
A82	Headrests (w/std bench front seat)	4,799	42.15
A85	Custom deluxe belts	2,957	26.35
A96	Stowage compartment lock (wagon only)	9,674	10.55
B02	Taxi equipment	1,355	56.90–62.15
B34	HD floor mats	9,257	9.50
B35	HD floor mats	8,479	9.50
B37	2 front & 2 rear color-keyed floor mats	132,865	10.55
B39	Load floor area carpet (wagons only)	5,021	52.70
B80	Bright roof drip molding	8,291	10.55–15.80
B90	Door & window frame moldings	7,697	21.10–26.35
B93	Door edge guards	196,934	3.20–6.35
C08	Vinyl roof cover	218,127	79.00
C48	Heater & defroster deletion	5,735	70.50 CR
C50	Rear window defroster	43,778	21.10
C51	Rear window air deflector (wagons only)	21,989	19.00
C60	Four Season AC	453,861	356.00
C75	Comfortron AC	27,370	435.00
D33	Remote control outside LH mirror	57,485	9.50
D96	Special bodyside accent stripes	200	—
F40	Special front & rear suspension	50,126	4.75–15.80

43

Code	Description	Quantity	Price
F41	Special purpose front & rear suspension	2,991	31.60
G66	Superlift rear shock absorbers	19,590	36.90-84.30
G75	3.70 ratio rear axle	372	2.15
G76	3.36 ratio rear axle	824	2.15
G80	Positraction rear axle	64,507	42.15
G92	3.08 ratio rear axle	1,025	2.15
G94	3.31 ratio rear axle	1,336	2.15
G96	3.55 ratio rear axle	5,099	2.15
G97	2.73 ratio rear axle	2,611	2.15
H01	3.07 ratio rear axle	823	2.15
H05	3.73 ratio rear axle	836	2.15
J50	Power brakes	343,284	42.15
J52	Front disc brakes	6,351	79.00
J65	Brakes w/metallic facings	1,588	36.90
K02	Radiator fan	6,183	15.80
K19	AIR equipment	83,027	44.75
K24	Closed positive engine ventilation	86,619	5.25
K30	Speed & cruise control	6,815	50.05
K45	HD oil bath air cleaner	5,020	5.30
K76	61 amp Delcotron generator	2,018	21.10
K79	42 amp Delcotron generator	3,138	10.55
K81	62 amp Delcotron generator	1,539	63.20-73.75
L30	275 hp Turbo-Fire 327 engine	340,347	92.70
L35	325 hp Turbo-Jet 396 engine	61,945	158.00
L36	385 hp Turbo-Jet 427 engine	4,337	316.00
M01	HD clutch	2,106	10.55
M10	Overdrive transmission	3,566	115.90
M13	Special 3 speed fully synchronized transmission	699	79.00
M20	4 speed wide-range transmission	14,600	184.35
M35	Powerglide transmission	907,679	184.35-194.85
M40	Turbo Hydra-matic transmission	155,292	226.45
N10	Dual exhaust	36,826	21.10
N30	Deluxe steering wheel	27,124	4.25
N33	Comfortilt steering wheel	48,886	42.15
N34	Sports-styled wood-grained steering wheel w/plastic rim	3,833	31.60
N40	Power steering	925,000	94.80

N96	Mag-style wheel covers	5,140	52.70-73.75
PQ8	8.25-14-4 pr white stripe tubeless tires	1,014	51.00
PU1	G70-15-4 pr white stripe tubeless tires	794	62.00-67.10
PU2	G70-15-4 pr red stripe tubeless tires	236	62.00
P01	Brightmetal wheel covers	565,658	21.10
P02	Simulated wire wheel covers	12,169	55.85-73.75
P12	5 14x6JK wheels	22,507	5.30
P42	5 15x6JK wheels	—	5.30
P85	8.55-14-4 pr whitewall tubeless tires	78,132	NC
QC2	8.45-15-8 pr whitewall tubeless tires (wagons)	748	Std
R51	8.15-15-4 pr whitewall tubeless tires	2,969	35.55
T58	Rear fender skirts	58,277	26.35
T60	HD battery	76,022	7.40
T78	Front fender lights	196,285	21.10
U03	Tri-Volume horn	6,356	13.70
U14	Special instrumentation	12,602	79.00
U15	Speed warning indicator	25,630	10.55
U25	Luggage compartment lamp	8,947	2.65
U26	Underhood lamp	89,048	2.65
U27	Glove compartment lamp	2,447	2.65
U28	Ashtray lamp	102,773	1.60
U29	Instrument panel courtesy lights	71,210	4.25
U35	Electric clock	8,847	15.80
U57	Stereo tape system	19,814	128.50
U63	Push-button control radio	919,935	57.40-70.60
U69	Push-button control AM/FM radio	32,069	133.80-147.00
U73	Manual rear antenna	125,134	9.50
U75	Power rear antenna	2,681	28.45
U79	Push-button control AM/FM stereo radio	10,598	239.15
U80	Rear seat speaker	199,623	13.20
V01	HD radiator	15,314	10.55
V31	Front bumper guard	146,255	15.80
V32	Rear bumper guard	131,038	15.80
V54	Luggage carrier (wagons only)	1,584	42.15
V55	Deluxe adjustable luggage carrier (wagons only)	45,177	63.20
Z04	HD chassis equipment	430	36.90
Z24	385 hp 427 ci SS 427 engine	2,124	403.30

894	Interior trim	277,087	5.30–10.55
950	Two-tone exterior paint combinations	54,958	15.80
—	Appearance Guard Group		32.70–54.85
—	Auxiliary Lighting Group		4.25–34.90
—	Decor Group		36.90–83.35
—	Foundation Group		80.60
—	Operation Convenience Group		30.60
—	Station Wagon Convenience Group		61.15–103.30

*Options include Biscayne, Bel Air, Impala and Caprice models.

Facts

The 1967 Impala was again restyled, looking longer and heavier, with the fastback roof design being even more pronounced. On the Super Sport models, the new front grille featured blacked-out horizontal strips, and on the rear was a black panel insert between the taillights. The SS identification could be found on the grille, trunk lid and each front fender. Black lower body and rear fender moldings were used on the Super Sport models in contrast with the bright pieces found on other Impalas. The standard Impala wheel covers were used with SS center caps.

In the interior, the usual vinyl buckets with center console were standard; however, a bench seat with a folding armrest could be substituted at no cost. The SS identification was limited to an SS emblem on the glovebox door. Brushed aluminum inserts embellished the console and dash panel. The dash came with three large round gauges arranged horizontally, with two auxiliary gauges on each end arranged vertically.

Engine availability consisted of the 250 ci for six-cylinder Impala SS models. The V-8 Super Sport Impalas came with the 283 ci single-exhaust two-barrel engine rated at 195 hp, a dual-exhaust four-barrel 327 rated at 275 hp and the 325 hp 395 ci Turbo-Jet.

Only one 427 ci V-8 was optional. This was the L36, now rated at 385 hp. On the Impala, it was part of the SS 427 option, a combination of the Z24 Super Sport option and the L36 engine option. The SS 427 came with a unique domed hood with three simulated air ducts, large SS 427 emblems in the center of the grille and on the rear panel between the taillights, heavy-duty suspension components and Red Line 8.25x14 tires on 14x6 in. rims. A total 2,124 Impala SS 427s were produced. A total 4,337 of the 427 engines were installed in all 1967 Chevrolet full-size models.

Optional for the first time were front disc brakes. Power brakes and 15 in. rally wheels were mandatory with the disc brakes.

The 1967 Impala SS Convertible.

Chapter 8

1968 Impala SS and SS 427

Production

2 dr coupe & convertible, all engines	36,432
2 dr coupe & Z24 convertible, 427 ci	1,778
Total	38,210

Serial numbers

Description
164678F100001
16467 — Model number (16367-2 dr convertible, 16387-2 dr coupe, 16467-2 dr convertible, 16487-2 dr coupe)
8 — Last digit of model year (1968)
F — Assembly plant (C-Southgate, D-Doraville, F-Flint, J-Janeville, L-Los Angeles, R-Arlington, S-St Louis, T-Tarrytown, U-Lordstown, Y-Wilmington, 2-St Therese)
100001 — Consecutive sequence number

Location
    On plate attached to driver's side of dash, visible through the windshield.

Engine and transmission suffix codes
CA, CJ, CM — 250 ci I-6 1 bbl 155 hp, 3 speed manual
CC, CK, CN — 250 ci I-6 1 bbl 155 hp, 3 speed manual w/AC
CQ — 250 ci I-6 1 bbl 155 hp, Powerglide automatic
CR — 250 ci I-6 1 bbl 155 hp, Powerglide automatic w/AC
DO, DQ — 307 ci V-8 2 bbl 200 hp, 3 speed manual
DP — 307 ci V-8 2 bbl 200 hp, 4 speed manual
DH, DR — 307 ci V-8 2 bbl 200 hp, Powerglide automatic
DK — 307 ci V-8 2 bbl 200 hp, Turbo Hydra-matic automatic TH400
DS — 307 ci V-8 2 bbl 200 hp, 3 speed automatic
HA, HB — 327 ci V-8 4 bbl 250 hp, 3 or 4 speed manual
HC, HG, HH — 327 ci V-8 4 bbl 250 hp, Powerglide automatic
HF — 327 ci V-8 4 bbl 250 hp, Turbo Hydra-matic automatic
HI, HL — 327 ci V-8 4 bbl 275 hp, 3 or 4 speed manual
HJ — 327 ci V-8 4 bbl 275 hp, Powerglide automatic
HM — 327 ci V-8 4 bbl 275 hp, Turbo Hydra-matic automatic
IA — 396 ci V-8 4 bbl 325 hp, 3 or 4 speed manual
IG — 396 ci V-8 4 bbl 325 hp, Powerglide automatic
IV — 396 ci V-8 4 bbl 325 hp, Turbo Hydra-matic automatic
IE, IH — 427 ci V-8 4 bbl 385 hp, 3 or 4 speed manual
IJ, IS — 427 ci V-8 4 bbl 385 hp, Turbo Hydra-matic automatic
IC — 427 ci V-8 4 bbl 385 hp, manual police

IB — 427 ci V-8 4 bbl 385 hp, Turbo Hydra-matic automatic police
ID — 427 ci V-8 4 bbl 425 hp, special high performance

Carburetors
396 ci — 7027201 manual, 7027200 automatic
427 ci 385 hp — 7027211 manual, 7027210 automatic
427 ci 425 hp — 3923289 (Holley R-4053A)

Distributors
396 ci — 1111169
427 ci 385 hp — 1111169

Exterior color codes

Tuxedo Black	AA
Ermine White	CC
Grotto Blue	DD
Fathom Blue	EE
Island Teal	FF
Ash Gold	GG
Grecian Green	HH
Tripoli Turquoise	KK
Teal Blue	LL
Cordovan Maroon	NN
Seafrost Green	PP
Matador Red	RR
Palomino Ivory	TT
Sequoia Green	VV
Butternut Yellow	YY

Convertible top color codes

White	AA
Black	BB
Blue	

Two-tone color codes*

Grotto Blue/Ermine White	DC
Grotto Blue/Fathom Blue	DE
Fathom Blue/Grotto Blue	ED
Ash Gold/Palomino Ivory	GT
Tripoli Turquoise/ Ermine White	KC
Teal Blue/Island Teal	LF

*Lower/upper.

Interior trim codes

Black	805/806/812/813
Blue	820/821
Gold	830/833/836
Saddle	839
Turquoise	842/845
Parchment	858/859
Teal	861/862/864
Red	866/868

Vinyl top color codes

Black	AA
White	

Options*

16447 Custom coupe	$2,998.00
16487 Sport coupe	2,945.00
16467 Convertible	3,197.00

Option number	Description	Quantity	Retail price
AS1	Seatbelts & shoulder belts	82	$ 23.20
AS4	2 front & 2 rear custom deluxe shoulder belts	472	52.70
AS5	2 front & 2 rear std-style shoulder belts	179	46.40
A01	Soft Ray tinted glass (all windows)	566,035	39.50

Code	Description	Qty	Price
A02	Soft Ray tinted glass (windshield only)	314,487	25.30
A31	Power windows	39,826	100.10
A33	Power tailgate window	78,771	31.60
A39	Custom deluxe seatbelts	4,841	7.90–12.65
A42	6 way power front seat	18,486	94.80
A46	4 way power seat	1,509	69.55
A51	Strato-Bucket seats	8,081	158.00
A53	Strato-Back seat	14,089	105.35
A75	HD seats	8,980	11.60–23.20
A76	HD rear seats	3,100	6.35
A81	Head restraints (w/Strato-Back or bucket front seats)	2,011	52.70
A82	Head restraints (w/std full-width front seat)	6,498	42.15
A85	Custom deluxe shoulder belts	429	26.35
A93	Power door lock system	7,413	44.80–68.50
A96	Rear compartment lock	5,495	10.55
B02	Taxi equipment	1,586	68.50–79.00
B07	Police car chassis equipment	7,687	68.50
B34	HD front floor mats	6,498	NC
B35	HD rear floor mats	5,732	NC
B37	2 front & 2 rear color-keyed floor mats	259,158	10.55
B39	Load floor carpet	8,346	52.70
B55	Extra-thick foam seat cushion	12,830	7.40
B80	Bright roof drip molding	4,061	10.55–15.80
B90	Door & window frame moldings	11,983	21.10–26.35
B93	Door edge guards	332,557	4.25–7.40
C05	Convertible tops	—	NC
C08	Vinyl roof cover	354,546	89.55
C51	Rear window air deflector	23,402	19.00
C56	Astro ventilation	7,585	15.80
C60	Four Season AC	590,633	368.65
C75	Comfortron AC	28,870	447.65
D33	Remote control outside rearview mirror	63,963	9.50
F40	Special front & rear suspension	37,410	4.75–15.80
F41	Special purpose front & rear suspension	3,092	21.10

G66	Superlift air-adjustable shock absorbers	23,154	42.15
G67	Superlift air-adjustable shock absorbers w/automatic level control	4,955	89.55
G75	3.70 ratio rear axle	424	2.15
G76	3.36 ratio rear axle	2,599	2.15
G80	Positraction rear axle	64,071	42.15
G92	3.08 ratio rear axle	1,171	2.15
G94	3.31 ratio rear axle	1,087	2.15
G96	3.55 ratio rear axle	3,616	2.15
G97	2.73 ratio rear axle	8,740	2.15
H01	3.07 ratio rear axle	1,355	2.15
H05	3.73 ratio rear axle	448	2.15
J50	Power brakes	411,715	42.15
J52	Power disc brakes	29,826	121.15
KD5	HD engine ventilation	131	6.35
K02	Temperature-controlled fan	4,236	15.80
K05	Engine block heater	1	—
K30	Cruise-Master speed control	15,564	52.70
K45	HD air cleaner	2,935	5.30
K76	61 amp Delcotron generator	7,200	5.30–26.35
K79	42 amp Delcotron generator (incl w/Turbo Hydra-matic)	2,797	10.55
L30	275 hp Turbo-Fire 327 V–8 engine	230,431	92.70
L35	325 hp Turbo-Jet 396 V–8 engine	55,190	158.00
L36	385 hp Turbo-Jet 427 V–8 engine (incl w/SS 427equipment)	4,071	263.30
L72	425 hp Turbo-Jet 427 V–8 engine (incl w/SS 427 equipment)	568	447.65
L73	250 hp Turbo-Fire 327 V–8 engine	277,431	63.20
M01	HD clutch	2,183	10.55
M10	Overdrive transmission	2,711	115.90
M13	Special 3 speed transmission	365	79.00
M20	4 speed wide-range transmission	6,596	184.35
M21	4 speed close-ratio transmission	1,052	184.35
M22	HD 4 speed close-ratio transmission	124	310.70

Code	Description	Quantity	Price
M35	Powerglide transmission	801,255	184.35–194.85
M40	Turbo Hydra-matic transmission	390,085	226.45–237.00
N10	Dual exhaust	39,631	27.40
N30	Deluxe steering wheel	54,424	4.25
N33	Comfortilt steering wheel	48,527	42.15
N34	Sports-styled steering wheel	2,021	31.60
N40	Power steering	1,082,395	94.80
N95	Simulated wire wheel covers	9,782	55.85–73.75
N96	Mag-style wheel covers	4,921	52.70–73.75
PA2	Mag-spoke wheel covers	428	52.70–73.75
PQ7	8.25–14–4 pr whitewall tubeless tires	4,578	51.00
PQ8	8.25–14–4 pr white stripe tubeless tires	675	51.00
PR3	8.25–14–8 pr whitewall tubeless tires	9,674	85.15
PS6	8.55–14–8 pr whitewall tubeless tires	3,766	87.05
PU1	G70x15 2 ply special white stripe tubeless tires	—	71.00–76.15
PU2	G70x15 2 ply special red stripe tubeless tires	—	71.00–76.15
PU3	G70x15–4 pr white stripe tubeless tires	3,124	—
PU4	G70x15–4 pr red stripe tubeless tires	1,029	—
P01	Wheel covers	660,273	21.10
P12	14x6JK wheels	13,170	5.30
P77	8.25–14–4 pr whitewall tubeless tires	793,293	35.55
P85	8.55–14–4 pr whitewall tubeless tires	104,760	35.50
P99	8.45–15–4 pr whitewall tubeless tires	5,442	66.95
Q04	8.25–14–4 pr blackwall tubeless tires	—	8.65
R51	8.15–15–4 pr whitewall tubeless tires	13,818	43.35
T58	Rear fender skirts	84,665	26.35
T60	HD 70 amp-hr battery	91,133	7.40
T83	Concealed headlights	14,929	79.00
U03	Tri-Volume horn	5,530	13.70
U14	Special instrumentation	8,858	79.00–94.80
U15	Speed warning indicator	23,596	10.55
U23	Ignition switch light	1,315	2.65

Code	Option	Quantity	Price
U25	Luggage compartment lamp	4,111	2.65
U26	Underhood lamp	123,852	2.65
U27	Glove compartment lamp	1,314	2.65
U28	Ashtray lamp	101,754	1.60
U29	Instrument panel courtesy lights	97,263	2.65
U35	Electric clock	307,264	15.80
U46	Light monitoring system	19,679	26.35
U57	Stereo tape system	17,783	133.80
U63	Push-button radio	1,026,914	61.10
U69	Push-button AM/FM radio	45,423	133.80
U73	Manual rear antenna	80,200	9.50
U75	Power rear antenna	4,107	28.45
U79	AM/FM stereo radio	12,450	239.15
U80	Auxiliary speaker	182,543	13.20
V01	HD radiator	16,327	13.70
V31	Front bumper guard	150,937	15.80
V32	Rear bumper guard	130,769	15.80
V54	Deluxe luggage carrier	2,369	63.20
V55	Luggage carrier (wagon)	62,017	44.25
ZJ7	Rally wheels	14,998	10.55–31.60
ZJ9	Auxiliary lighting	106,721	2.65–39.00
Z03	Impala SS equipment	36,432	179.05
Z04	HD chassis equipment	259	36.90
Z24	SS 427 equipment	1,778	358.10–542.45
865	All-vinyl interior trim	404,029	5.30–10.55
950	Two-tone paint combinations	67,585	—

*Options are for all Chevrolet Biscayne, Bel Air, Impala and Caprice models.

Facts

Reflecting the increased emphasis on luxury, Super Sport Impalas were downgraded to an option package, RPO Z03, rather than a separate model line. The SS 427 Package continued as RPO Z24, with the 385 hp 427 ci engine, but a 425 hp 427 ci was optionally available. This was the L72 427 with the big-port heads, aluminum intake and Holley 780 cfm combination, 11.0:1 compression and solid-lifter camshaft. Transmission availability was either a four-speed manual or a Turbo Hydra-matic automatic.

In addition to the minor facelift on the 1968 Impala, new styling features included hideaway windshield wipers and front and rear side-mounted marker lights, which were federally mandated. On the rear, the Impala returned to the three-per-side taillight motif, but these were mounted within the new bumper. In the interior, the dash panel was again redesigned, reverting to use of the wide-band speedometer.

From 1968, the VIN plate was attached on the left side of the dash and was visible through the windshield—another government-mandated change.

The SS Package consisted of front grille, rear deck and front fender identification, and full wheel covers. In the interior, bucket seats were standard equipment, with a center console.

The SS 427 added a blacked-out grille, a modified hood with nonfunctional air intakes, three louvers on each front fender and SS 427 identification on the center of the grille and on the right side of the trunk lid. A set of G70x15 Red Line tires was standard. In the engine compartment, the 427 got a chrome air cleaner, valve covers and oil filler cap. As before, the 427s came with heavy-duty suspension.

A total 568 of the L72 425 hp powered Impalas were built, in addition to 4,071 of the L36 385 hp cars. The SS 427 cars are included in these figures.

From 1968 on, Chevrolet switched from the previous canister setup to a spin-on oil filter.

The 1968 Impala SS Convertible.

The 1969 Impala SS 427 Convertible.

Chapter 9

1969 Impala SS 427

Production
RPO Z24 SS 427   2,455

Serial numbers

Description
164379F100001
16437 — Model number (16437-2 dr coupe, 16447-2 dr coupe, 16467-2 dr convertible)
9 — Last digit of model year (1969)
F — Assembly plant (C-Southgate, D-Doraville, F-Flint, J-Janeville, L-Los Angeles, R-Arlington, S-St Louis, T-Tarrytown, U-Lordstown, Y-Wilmington, 2-St Therese)
100001 — Consecutive sequence number

Location
On plate attached to driver's side of dash, visible through the windshield.

Engine and transmission suffix codes
LA, LH, MA, MC — 427 ci V-8 4 bbl 390 hp, manual
LC, LE, LI — 427 ci V-8 4 bbl 390 hp, Turbo Hydra-matic automatic TH400
LD, MD — 427 ci V-8 4 bbl 425 hp, manual
LS — 427 ci V-8 4 bbl 425 hp, Turbo Hydra-matic automatic TH400

Carburetors
427 ci 390 hp — 7029201 manual, 7029200 automatic
427 ci 425 hp — 3959164 Holley R 4346A

Distributors
427 ci 390 hp — 1111925

Exterior color codes

Tuxedo Black	10	Fathom Green	57
Butternut Yellow	40	Frost Green	59
Dover White	50	Burnished Brown	61
Dusk Blue	51	Champagne	63
Garnet Red	52	Olympic Gold	65
Glacier Blue	53	Burgundy	67
Azure Turquoise	55	Cortez Silver	69

Two-tone color codes*

Glacier Blue/ Dover White	53/50	Glacier Blue/Dusk Blue	53/51
		Dusk Blue/Glacier White	51/53

Two-tone color codes*

Olympic Gold/ Dover White	65/50
Burnished Brown/ Champagne	61/63

Azure Turquoise/ Dover White	55/50
*Lower/upper.	

Interior trim codes

Black	805
Blue	820
Saddle	830
Gold	837
Turquoise	844
Green	852/854
Parchment	858/859
Midnight Green	861
Red	866/867

Convertible top color codes

White	AA
Black	BB

Vinyl top color codes

Black	BB
Parchment	EE
Dark Blue	CC
Dark Brown	FF
Midnight Green	GG

Options

16437 Sport coupe	$3,016.00
16447 Custom coupe	3,068.00
16467 Convertible	3,244.00

Option number	Description	Quantity	Retail price
AR1	Less head restraint	4,432	
AS1	Std shoulder harness	56	$ 23.20
AS4	Deluxe rear seat shoulder harness	314	26.35
AS5	Std rear seat shoulder harness	285	23.20
A01	Tinted glass (all windows)	803,482	42.15
A02	Tinted glass (windshield)	42,461	
A31	Electric control windows	51,727	105.35
A33	Electric tailgate window	74,976	34.80
A39	Custom deluxe front & rear seatbelts	2,328	9.00– 10.55
A42	6 way electric control front seat	22,500	100.10
A51	Strato-type front bucket seats	28,894	168.55
A53	Strato-type front bench seat	6,531	115.90
A75	HD low-profile front seat	8,140	12.65– 25.30
A76	HD rear seat	2,655	6.35
A85	Deluxe shoulder harness	352	26.35
A91	Vacuum control rear compartment lid release	16,177	14.75
A93	Vacuum-operated door locks	17,106	44.80– 68.50
A96	Rear compartment lock	5,692	—
B02	Taxicab equipment	1,701	—

B07	Police car equipment	7,275	—
B34	HD front floor mat	5,103	2.15
B35	HD rear floor mat	4,087	NC
B37	Floor mats	270,812	11.60
B39	Load floor carpet	13,454	—
B55	Deluxe front seat cushion	11,746	7.40
B80	Roof drip molding	8,047	11.60
B84	Bodyside molding	2,302	26.35
B90	Door & window frame molding	7,601	21.10
B93	Door edge guards	348,764	4.25–7.40
CE1	Headlamp washer	1,540	15.80
C08	Exterior soft trim roof cover	435,542	88.55
C50	Rear window defroster	73,996	22.15–32.65
C51	Rear window air deflector	38,670	—
C60	Deluxe AC	700,038	384.45
C75	Automatic temperature control AC	30,396	463.45
D33	Remote control outside mirror	89,863	10.55
D34	Vanity Visor mirror	80,630	3.20
F40	Special front & rear suspension	41,585	5.30–16.90
F41	Special performance front & rear suspension	1,572	22.15
GT1	2.56 ratio rear axle	11,291	2.15
GT2	2.29 ratio rear axle	511	2.15
G66	Air booster rear shock absorber	27,329	42.15
G67	Rear shock absorber level control	5,546	89.55
G76	3.36 ratio rear axle	927	2.15
G80	Positraction rear axle	58,082	42.15
G92	3.08 ratio rear axle	1,221	2.15
G94	3.31 ratio rear axle	1,957	2.15
G96	3.55 ratio rear axle	1,687	2.15
G97	2.73 ratio rear axle	3,573	2.15
H01	3.07 ratio rear axle	8,257	2.15
H05	3.73 ratio rear axle	380	2.15
J50	Vacuum power brake equipment	492,373	42.15
J52	Power front disc brakes	125,334	64.25
KD5	HD closed positive ventilation	327	6.35
K02	Fan drive equipment	3,239	15.80
K05	Engine block heater	11,054	10.55
K30	Speed & cruise control	20,421	57.95
K45	Oil bath air cleaner (1 lb capacity)	2,576	5.30

Code	Description	Qty	Price
K79	42 amp AC generator	2,553	10.55
K85	63 amp AC generator	10,159	5.30–26.35
LM1	350 ci V-8 engine (regular fuel)	217,232	52.70
LS1	427 ci V-8 engine	18,308	163.25
L36	427 ci Hi-Performance V-8 engine	5,582	237.00
L48	350 ci V-8 engine	211,073	52.70
L65	350 ci 2 bbl V-8 engine	18	21.10
L66	396 ci 2 bbl V-8 engine	63,603	68.50
L72	427 ci Special Hi-Performance V-8 engine	546	447.65
MC1	HD 3 speed transmission	1,926	79.00
M20	4 speed transmission	3,448	184.80
M21	4 speed close-ratio transmission	969	184.80
M22	HD 4 speed transmission	77	313.00
M35	Powerglide transmission	536,651	163.70–174.25
M38	300 Deluxe 3 speed automatic transmission	492,773	—
M40	3 speed automatic transmission	111,241	190.10–221.80
N10	Dual exhaust system	32,646	30.55
N33	Tilt-type steering wheel	73,155	45.30
N34	Wood-grain plastic steering wheel	2,215	34.80
N40	Power steering	1,078,185	100.10–105.35
N95	Simulated wire wheel trim cover	5,494	55.85–73.75
N96	Simulated magnesium wheel trim cover type A	2,608	52.70–73.75
PA2	Simulated magnesium wheel trim cover type B	749	52.70–73.75
PK2	G78-14-4 pr B/B whitewall tires	12,637	73.80
PQ7	8.25-14-4 pr whitewall tires	2,619	51.00
PR3	8.25-14-4 pr whitewall tires	5,965	85.10
PS6	8.55-14-8 pr whitewall tires	1,238	—
PS8	8.55-14-4 pr whitewall tires	75,862	—
PU3	G70-15-4 pr white strip tires	3,351	63.40–77.05
PU4	G70-15-4 pr red stripe tires	2,188	63.40
PU8	G78-15-4 pr B/B whitewall tires	13,492	—

P01	Wheel trim cover	721,934	21.10
P02	Deluxe wheel trim cover	9,294	57.95–79.00
P06	Wheel trim ring	1,027	21.10
P12	14x6JK wheel equipment	8,211	5.30
P74	8.25-14-4 pr whitewall dual-stripe tires	71,986	40.75
P77	8.25-14-4 pr whitewall tires	534,549	35.50
P85	8.55-14-4 pr whitewall tires	3,561	—
P89	8.55-14-8 pr whitewall tires	6,907	—
P90	G70-15-4 pr white stripe tires	6,944	25.70–86.90
P91	G70-15-4 pr B/B red stripe tires	640	25.70–86.90
P99	8.45-15-4 pr whitewall tires	21	—
QA7	8.25-15-4 pr whitewall dual-stripe tires	26,811	40.75–54.40
QC2	8.45-15-8 pr whitewall tires	29,181	—
R51	8.15-15-4 pr whitewall tires	47,134	35.55
T58	Rear wheel opening skirt	71,270	
T60	HD battery	100,334	8.45–15.80
T83	Retractable headlamp cover	23,797	52.70
UF1	Map light	81,587	
U15	Speed warning indicator	22,463	11.60
U25	Luggage compartment lamp	3,691	7.90–17.40
U26	Underhood lamp	81,564	7.90–17.40
U27	Glove compartment lamp	1,823	7.90–17.40
U28	Ashtray lamp	3,533	7.90–17.40
U29	Instrument panel courtesy lights	50,796	7.90–17.40
U35	Electric clock	256,962	15.80
U46	Lamp monitoring	18,569	26.35
U57	Tape player	21,099	133.80
U63	Push-button radio	982,579	61.10
U69	Push-button AM/FM radio	45,446	133.80
U73	Manual rear antenna	59,816	10.55
U79	Stereo equipment	17,509	239.10
U80	Auxiliary speaker	192,516	13.20
V01	HD radiator	12,606	14.75

V31	Front bumper guard equipment	94,989	15.80
V32	Rear bumper guard equipment	77,085	15.80
V55	Luggage carrier equipment	78,183	52.70
V75	Traction compound & dispenser	2,086	23.20
ZJ4	Seatbelt, check-doors & low-fuel warning lights	34,774	—
ZJ7	Special wheel, hubcap & trim ring	27,912	21.10-35.85
ZJ9	Auxiliary Lighting Group	79,541	7.90-17.40
ZK1	Boby Insulation Package	39,169	—
ZK3	Deluxe seatbelts & front seat shoulder harness	185,661	12.15-13.70
Z21	Exterior molding	18,877	36.90-47.40
Z24	SS 427 equipment	2,455	422.35
867	Vinyl-coated trim	338,503	6.35-12.65
959	Two-tone color combinations	110,777	23.20

Facts

The year 1969 marked the end of the Super Sport Impala. For this year, the series was limited to just the SS 427 option, with either the 390 hp 427—uprated 5 hp from the 1968 version—or the 425 hp solid-lifter 427.

The L36 390 hp 427 was available on all other Impala models; 5,582 were so equipped, including the SS 427s. Only 546 Impalas got the 425 hp 427.

The SS identification was limited to an SS emblem on the steering wheel in the interior. Bucket seats and console, which had been standard equipment with the Super Sport, were now optional.

Exterior identification was limited to a blacked-out grille with SS lettering in the center, and fender and trunk lid emblems. Red Line G70x15 tires, power front disc brakes and heavy-duty suspension were standard.

Impala disc brakes came with single piston calipers.

The Impala itself got another restyle in 1969. Side vent windows were eliminated, the loop-type front bumber was all-new and the overall body design resembled the 1968's—it was fuller, giving the car a heavier, larger look. The rear bumper housed three rectangular taillights per side.

Chapter 10

1964 Chevelle Malibu SS

Production

5737 2 dr coupe, 6 cyl	8,224	5867 2 dr convertible, 8 cyl	9,640
5767 2 dr convertible, 6 cyl	1,551	Total	76,860
5837 2 dr coupe, 8 cyl	57,445		

Serial numbers

Description

45867A100001

4 — Last digit of model year (1964)

5867 — Model number (5737-2 dr coupe, 5767-2 dr convertible, 5837-2 dr coupe, 5867-2 dr convertible)

A — Assembly plant (A-Atlanta, B-Baltimore, H-Fremont, K-Kansas City, L-Los Angeles)

100001 — Consecutive sequence number

Location

On plate attached to left front door hinge post.

Engine and transmission suffix codes

G, GB — 194 ci I-6 1 bbl 120 hp, 3 speed manual
GF, GG — 194 ci I-6 1 bbl 120 hp, 3 speed manual w/PCV
GK, GL, GN — 194 ci I-6 1 bbl 120 hp, 3 speed manual w/AC
GM — 194 ci I-6 1 bbl 120 hp, 3 speed manual w/AC & PCV
K — 194 ci I-6 1 bbl 120 hp, Powerglide automatic
KB — 194 ci I-6 1 bbl 120 hp, Powerglide automatic w/PCV
KH — 194 ci I-6 1 bbl 120 hp, Powerglide automatic w/AC
KJ — 194 ci I-6 1 bbl 120 hp, Powerglide automatic w/AC & PCV
LM — 230 ci I-6 1 bbl 140 hp, 3 speed manual
LL — 230 ci I-6 1 bbl 140 hp, 3 speed manual w/AC
LN — 230 ci I-6 1 bbl 140 hp, 3 speed manual w/AC & PCV
BL — 230 ci I-6 1 bbl 140 hp, Powerglide automatic
BN — 230 ci I-6 1 bbl 140 hp, Powerglide automatic w/PCV
BM — 230 ci I-6 1 bbl 140 hp, Powerglide automatic w/AC
BP — 230 ci I-6 1 bbl 140 hp, Powerglide automatic w/AC & PCV
J — 283 ci V-8 2 bbl 195 hp, 3 speed manual
JA — 283 ci V-8 2 bbl 195 hp, 4 speed manual
JD — 283 ci V-8 2 bbl 195 hp, Powerglide automatic
JH — 283 ci V-8 4 bbl 220 hp, 3 or 4 speed manual
JG — 283 ci V-8 4 bbl 220 hp, Powerglide automatic
JQ — 327 ci V-8 4 bbl 250 hp, 3 or 4 speed manual

Engine and transmission suffix codes

JT — 327 ci V-8 4 bbl 250 hp, 3 or 4 speed manual w/transistor ignition
SR — 327 ci V-8 4 bbl 250 hp, Powerglide automatic
JR — 327 ci V-8 4 bbl 300 hp, 3 or 4 speed manual
SS — 327 ci V-8 4 bbl 300 hp, Powerglide automatic
JS — 327 ci V-8 4 bbl 300 hp, 4 speed manual, special high performance*
  *Horsepower not available.

Carburetors

327 ci 250 hp — 7016417
327 ci 250 hp w/Powerglide — 7016422
327 ci 300 hp — 3851761
327 ci 300 hp w/Powerglide — 3851762

Distributor

327 ci 4 bbl — 1111016

Exterior color codes

Tuxedo Black	900
Meadow Green	905
Bahama Green	908
Silver Blue	912
Daytona Blue	916
Azure Aqua	918
Lagoon Aqua	919
Almond Fawn	920
Ember Red	922
Saddle Tan	932
Ermine White	936
Desert Beige	938
Satin Silver	940
Goldwood Yellow	943
Palomar Red	948

Interior trim codes

Saddle	710
Black	714
Aqua	722
Blue	741
Fawn	770
Red	786
White/Red	729

Convertible top color codes

White	Std
Beige	AB
Black	AA

Options*

5737 2 dr sport coue, 6 cyl	$2,538.00
5767 2 dr convertible, 6 cyl	2,749.00
5837 2 dr sport coupe, 8 cyl	2,646.00
5867 2 dr convertible, 8 cyl	2,857.00

Option number	Description	Quantity	Retail price
A01	Soft Ray tinted glass (all windows)	49,148	$ 31.25
A02	Soft Ray tinted glass (windshield only)	134,682	19.95
A31	Power windows (exc model 5315)	2,494	59.20–102.25
A33	Power rear window (for 2 seat station wagons only; std on 3 seat models)	10,240	26.90

A41	4 way electric control power seat (front seat only; NA on Malibu SS)	874	64.60
A49	Custom deluxe front seatbelts w/retractors (driver & passenger)	58,117	7.55
A62	Front seatbelts deletion (driver & passenger)	—	11.00 CR
A66	Divided second seat (for station wagons; Fawn trim only)	1,282	37.70
B01	HD body equipment (model 5369 only)	19	18.30
B02	Taxi equipment	101	63.50
B70	Padded instrument panel	143,161	18.30
C05	Convertible tops (models 5567 & 5767 only; choice of colors)	9,843	NC
C06	Power top (models 5567 & 5767 only)	9,415	53.80
C48	Heater & defroster deletion (NA w/AC)	13,656	72.00 CR
C50	Rear window defroster (sedans & sport coupes only)	—	21.55
C60	Four Season AC (incl 55 amp Delcotron & HD radiator; 7.00-14 or larger tires required)	31,861	363.70
C65	Custom deluxe AC (incl 55 amp Delcotron; 7.00-14 or larger tires required)	1,088	317.45
F40	Special front & rear suspension	22,115	
	On sedans, sport coupes & convertible, exc SS models, incl special front & rear springs & special front & rear shock absorbers		4.85
	On station wagons & Super Sport models, incl special front & rear springs		3.80
G66	Superlift rear shock absorbers	887	37.70
G76	3.36 ratio rear axle (incl when AC is ordered)	9,212	2.20
G80	Positraction rear axle	29,501	
	3.08 ratio		37.70
	3.36 ratio		37.70
	3.70 ratio		37.70

Code	Description	Qty	Price
J50	Vacuum brakes	37,739	43.05
J65	Special brakes w/metallic facings	1,573	37.70
K02	Temperature-controlled radiator fan	368	16.15
K24	Closed positive engine ventilation type B (approved by state of Calif)	36,934	5.40
K77	55 amp Delcotron generator (NA w/taxi equipment; incl w/AC)	280	21.55
K79	42 amp Delcotron generator	715	10.80
K81	62 amp Delcotron generator	99	
	For use wo/AC		75.35
	For use w/AC		64.60
L30	250 hp Turbo-Fire 327 V-8 engine	6,598	94.70
L61	155 hp Turbo-Thrift 230 engine	64,543	43.05
L77	220 hp Turbo-Fire 283 V-8 engine	54,840	53.80
M01	HD clutch (incl when taxi or HD chassis equipment is ordered)	673	5.40
M10	Overdrive transmission	5,241	107.60
M20	4 speed synchromesh transmission	30,566	188.30
M35	Powerglide transmission	241,412	199.10
N33	7 position Comfortilt steering wheel (power steering w/Powerglide or 4 speed transmission required)	3,646	43.05
N34	Sports-styled walnut-grained steering wheel w/plastic ring	3,131	32.30
N40	Power steering	111,590	86.10
P01	4 brightmetal wheel covers (NA on Malibu SS series)	125,480	18.30
P02	Simulated wire wheel covers	8,040	57.05- 75.35
T60	HD 66 plate 70 amp-hr battery	11,114	7.55
U16	Tachometer	5,734	48.45
U60	Manual control radio	52,088	50.05
U63	Push-button control radio	138,424	58.65
V01	HD radiator (NA w/AC)	24.492	10.80
V31	Front bumper guard	20,890	9.70

V32	Rear bumper guard (exc station wagons)	14,251	9.70
V55	Luggage carrier (station wagons only)	9,575	43.05
Z01	Comfort & convenience equipment type A	186,728	30.15–40.90
Z02	Push-button control radio & rear speaker	20,696	72.10
Z04	HD chassis equipment (NA w/AC or 155 hp engine; incl HD front & rear springs & HD front & rear shock absorbers)	27	19.40
Z13	Comfort & convenience equipment type B	10,503	39.85–50.60
759	Vinyl interior trim		107.60
	Exterior paint		
	Single colors		NC
	Two-tone combinations		16.15

*Options are for all Chevelle models.

Facts

The Chevelle was Chevrolet's midsize entry for 1964. Totally new, it resembled a small Impala. The optional SS Package followed the theme set by the Impala's Super Sport option.

The Super Sport was available only on the two-door coupe or convertible with any regular production engine. Six-cylinder Chevelles got a different model and series number than did V-8 equipped cars. The SS identification was found on the rear fenders, deck lid, door panels and glovebox door. Super Sport models also got sill moldings, lower rear fender moldings, bodyside moldings and wheelwell moldings. The Chevelle used the same full-size 14 in. wheel covers from the Impala.

In the interior, all Super Sport optioned cars got vinyl bucket seats. A console was used provided the car was optioned out with either the four-speed manual or Powerglide automatic. On the dash, a four-gauge cluster replaced the stock warning lights.

Midyear introductions were the 250 hp and 300 hp 327 ci small-block V-8s.

The 1964 Chevelle Malibu SS Coupe.

Chapter 11

1965 Chevelle Malibu SS and Chevelle SS 396

Production

13737 2 dr coupe, 6 cyl	7,452	13867 2 dr convertible,	
13767 2 dr convertible,		8 cyl	7,995
6 cyl	1,133	Total	81,112
13837 2 dr coupe, 8 cyl	64,532		

Serial numbers

Description

138675A100001

13867 — Model number (13737-2 dr coupe, 13767-2 dr convertible, 13837-2 dr coupe, 13867-2 dr convertible)

5 — Last digit of model year (1965)

A — Assembly plant (A-Atlanta, B-Baltimore, G-Framingham, K-Kansas City, Z-Fremont)

100001 — Consecutive sequence number

Location

On plate attached to left front door hinge post.

Engine and transmission suffix codes

AA, AC — 194 ci I-6 1 bbl 120 hp, 3 speed manual
AG, AH — 194 ci I-6 1 bbl 120 hp, 3 speed manual w/AC
AL — 194 ci I-6 1 bbl 120 hp, Powerglide automatic
AR — 194 ci I-6 1 bbl 120 hp, Powerglide automatic w/AC
CA — 230 ci I-6 1 bbl 140 hp, 3 speed manual
CB — 230 ci I-6 1 bbl 140 hp, 3 speed manual w/AC
CC — 230 ci I-6 1 bbl 140 hp, Powerglide automatic
CD — 230 ci I-6 1 bbl 140 hp, Powerglide automatic w/AC
DA — 283 ci V-8 2 bbl 195 hp, 3 speed manual
DB — 283 ci V-8 2 bbl 195 hp, 4 speed manual
DE — 283 ci V-8 2 bbl 195 hp, Powerglide automatic
DG — 283 ci V-8 4 bbl 220 hp, 3 speed manual
DH — 283 ci V-8 4 bbl 220 hp, Powerglide automatic
EA — 327 ci V-8 4 bbl 250 hp, 3 or 4 speed manual
EE — 327 ci V-8 4 bbl 250 hp, Powerglide automatic
EB — 327 ci V-8 4 bbl 300 hp, 3 or 4 speed manual
EF — 327 ci V-8 4 bbl 300 hp, Powerglide automatic
ED — 327 ci V-8 4 bbl 300 hp, transistor ignition
EC — 327 ci V-8 4 bbl 350 hp, 4 speed manual
IX — 396 ci V-8 4 bbl 375 hp, 4 speed manual

Carburetors

327 ci 250 hp — 7016418
327 ci 250 hp w/Powerglide — 7016516, 7023408
327 ci 300 hp — 3851761, 3862093, 3857360
327 ci 300 hp w/Powerglide — 3851762, 3862092, 3857359
327 ci 350 hp — 3863153, 3863150
396 ci — 3874898 (Holley R3139A), 3868864 (Holley R3140A, R3140-1AAS)

Distributors

327 ci — 1111075
327 ci 350 hp — 1111071
327 ci 350 hp w/transistor
  ignition — 1111072
396 ci — 1111100

Exterior color codes

Tuxedo Black	A
Ermine White	C
Mist Blue	D
Danube Blue	E
Willow Green	H
Cypress Green	J
Artesian Turquoise	K
Tahitian Turquoise	L
Madeira Maroon	N
Evening Orchid	P
Regal Red	R
Sierra Tan	S
Cameo Beige	V
Glacier Gray	W
Crocus Yellow	Y

Cylinder head casting number

396 ci — 3856208

Interior trim codes

Saddle	710
Black	714
Blue	741
Fawn	770
Red	786
Ivory/Black	792
Slate	799
Ivory/Aqua	796

Convertible top color codes

White	Std
Black	AA
Beige	AB

Vinyl top color code

Black	6

Options*

13737 2 dr sport coupe, 6 cyl	$2,539.00
13837 2 dr sport coupe, 8 cyl	2,647.00
13767 2 dr convertible, 6 cyl	2,750.00
13867 2 dr convertible, 8 cyl	2,858.00

Option number	Description	Quantity	Retail price
A01	Soft Ray tinted glass (all windows)	53,604	$ 31.25
A02	Soft Ray tinted glass (windshield only)	147,435	19.95
A33	Electric control power windows	2,052	102.25
A33	Power rear window	11,587	26.90
A41	4 way electric control power seat	717	64.60

A46	Electric control driver's bucket seat	23	NA
A47	Custom deluxe rear seatbelts	820	12.95
A49	Custom deluxe belts w/retractors	107,978	7.55
A62	Seatbelt deletion	29,864	11.00 CR
A66	Divided second seat	78	37.70
B70	Padded instrument panel	140,610	18.30
C06	Power top	8,710	53.80
C08	Black vinyl roof cover	2,442	75.35
C48	Heater & defroster deletion	7,751	72.00 CR
C50	Rear window defroster	1,667	21.55
C60	Four Season AC	42,098	363.70
F40	Special front & rear suspension	26,386	3.80–4.85
G66	Superlift rear shock absorbers	1,961	37.70
G67	Superlift rear shock absorbers w/automatic level control	587	86.15
G75	3.70 ratio rear axle	96	2.20
G76	3.36 ratio rear axle	2,759	2.20
G80	3.08, 3.36 & 3.70 ratio Positraction rear axles	23,023	37.70
J50	Vacuum power brakes	29,236	43.05
J65	Special brakes w/metallic facings	1,506	37.70
K02	Temperature-controlled radiator fan	343	16.15
K24	Closed positive engine ventilation	53,054	5.40
K66	Transistorized ignition system	209	75.35
K77	55 amp Delcotron generator	346	21.55
K79	42 amp Delcotron generator	776	10.80
K81	62 amp Delcotron generator	164	64.60–75.35
L26	140 hp Turbo-Thrift engine	68,741	26.90
L30	250 hp Turbo-Fire engine	36,261	94.70
L74	300 hp Turbo-Fire engine	13,593	137.75
L79	350 hp Turbo-Fire engine	6,021	202.30
M01	HD clutch	662	5.40
M10	Overdrive transmission	4,076	107.60
M20	4 speed synchromesh transmission	39,092	188.30
M35	Powerglide transmission	254,750	188.30–199.10
M55	Transmission oil cooler	113	16.15
N10	Dual exhaust	6,881	21.55

N33	Comfortilt steering wheel	7,251	43.05
N34	Sports-styled walnut-grained steering wheel w/plastic rim	2,166	32.30
N40	Power steering	125,808	86.10
P01	4 brightmetal wheel covers	125,202	21.55
P02	Simulated wire wheel covers	6,544	57.05–75.35
P19	Spare wheel lock	264	5.40
P58	7.35-14-4 pr whitewall tubeless tires	118,716	39.70
P61	7.75-14-4 pr nylon whitewall tubeless tires	2,391	54.10–73.20
P67	6.95-14-4 pr whitewall tubeless tires	93,406	28.70
T60	HD 66 plate 70 amp-hr battery	14,252	7.55
U03	Tri-Volume horn	5,461	14.00
U16	Tachometer	7,724	48.45
U60	Manual control radio	52,851	50.05
U63	Push-button control radio	187,895	58.65
U69	Push-button control AM/FM radio	2,794	136.70–150.15
U73	Rear antenna	2,794	NC
U80	Auxiliary speaker	24,658	
V01	HD radiator	21,104	10.80
V31	Front bumper guard	15,200	19.95
V32	Rear bumper guard	11,037	9.70
V55	Luggage carrier	7,423	43.05
Z01	Comfort & convenience equipment type A	209,641	30.15–40.90
Z13	Comfort & convenience equipment type B	7,405	39.85–50.60
Z16	Special sport coupe & convertible equipment	201	1,501.05
724	Bucket seats trim	3,234	NA
759	Vinyl interior trim	1,562	5.40
950	Two-tone exterior paint combinations	5,100	16.15

*Options are for all Chevelle models as of March 1, 1965.

Facts

The 1965 Chevelle was restyled, giving the car a longer, lower look. In the front, a new bumper and grille design complemented the slightly redesigned rear.

The SS identification consisted of SS rear fender emblems, an SS rear deck lid emblem, rocker panel moldings, a black-accented grille and a black band encircling the taillights and taillight panel. The 1965 Malibu SSs came with a full 14 in. wheel cover. Engine size identification was located on the front fenders.

In the interior, vinyl bucket seats were once again standard, as was the center console with four-speed- or Powerglide-equipped cars. As SS emblem was used on the glovebox door.

A higher-performance version of the 327 ci V-8 joined the line-up, rated at 350 hp and available only with a four-speed manual transmission. A total 6,021 Chevelles got this engine.

A midyear introduction was the Z16 Option Package costing $1,501.05. Its main claim to fame was the 396 ci big-block Mark IV engine rated at 375 hp. The 396 came with the big-port cylinder heads, 2.19 in. intake valves and 1.72 in. exhaust valves, aluminum intake manifold, Holley carburetor and, for 1965 only, a hydraulic-lifter camshaft. In addition to the usual SS identification, the Z16s got 396 Turbo-Jet insignias on both front fenders and rear deck lid. Tires were 7.75x14 gold stripe Firestones on 14 in. rims, which came with mag-style wheel covers. Also part of the package were heavy-duty suspension components that included a rear stabilizer bar. Brakes were borrowed from the full-size line, measured 11 in. and used Power Assist. In the interior were a 160 mph speedometer, 6000 rpm tachometer and custom accessory clock. Radio was an AM/FM stereo unit. An SS 396 emblem was mounted on the right side of the dash. A total 201 units were built: 200 coupes and one convertible.

The 1965 Chevelle Malibu SS Coupe.

Chapter 12

1966 Chevelle Malibu SS 396

Production

By engine
2 dr coupe & convertible,
  L35 396 ci V-8   44,362
2 dr coupe & convertible,
  L34 396 ci V-8   24,811
2 dr coupe & convertible,
  L78 396 ci V-8   3,099
  Total   72,272

By body style
13817 2 dr coupe, 8 cyl 66,843
13867 2 dr convertible,
  8 cyl   5,429
  Total   72,272

Serial numbers

Description
138176A100001
13817 — Model number (13817-2 dr coupe, 13867-convertible)
6 — Last digit of model year (1966)
A — Assembly plant (A-Atlanta, B-Baltimore, F-Flint,
  G-Framingham, K-Kansas City, Z-Fremont)
100001 — Consecutive sequence number

Location
On plate attached to left front door hinge post.

Engine and transmission suffix codes
ED — 396 ci V-8 4 bbl 325 hp, 3 or 4 speed manual
EH — 396 ci V-8 4 bbl 325 hp, 3 or 4 speed manual w/AIR
EK — 396 ci V-8 4 bbl 325 hp, Powerglide automatic
EM — 396 ci V-8 4 bbl 325 hp, Powerglide automatic w/AIR
EF — 396 ci V-8 4 bbl 360 hp, 3 or 4 speed manual
EJ — 396 ci V-8 4 bbl 360 hp, 3 or 4 speed manual w/AIR
EL — 396 ci V-8 4 bbl 360 hp, Powerglide automatic
EN — 396 ci V-8 4 bbl 360 hp, Powerglide automatic w/AIR
EG — 396 ci V-8 4 bbl 375 hp, 4 speed manual

Carburetors
396 ci 325 hp — 7016621
396 ci 325 hp — 3874898 (Holley R-3139-1AAS), engine ED
396 ci 325 hp w/Powerglide — 3868864 (Holley R-3140-1AAS),
  engine EK
396 ci 325 hp w/Powerglide — 7016620
396 ci 325 hp w/AIR — 7036201

396 ci 325 hp w/AIR & Powerglide — 7036200
396 ci 360 hp — 3886087 (Holley R-3419AAS)
396 ci 360 hp w/Powerglide — 3886088 (Holley R-3420A)
396 ci 360 hp w/AIR — 3886089 (Holley R-3421A)
396 ci 360 hp w/AIR — 3892339 (Holley R-3609A)
396 ci 360 hp w/AIR & Powerglide — 3892338 (Holley R-3608A)
396 ci 375 hp — 3893229 (Holley R-3613A)

Distributors
396 ci 325 hp — 1111109
396 ci 325 hp w/transistor
  ignition — 1111137
396 ci 360 hp — 1111138
396 ci 360 hp w/transistor
  ignition — 1111139
396 ci 375 hp — 1111100
396 ci 375 hp w/transistor
  ignition — 1111074

Cylinder head casting numbers
396 ci 325/360 hp — 3872702
396 ci 375 hp — 3873858

Vinyl top color codes
Black	BB
Beige	—

Exterior color codes
Tuxedo Black	A
Ermine White	C
Mist Blue	D
Danube Blue	E
Marina Blue	F
Willow Green	H
Artesian Turquoise	K
Tropic Turquoise	L
Aztec Bronze	M
Madeira Maroon	N
Regal Red	R
Sandalwood Tan	T
Cameo Beige	V
Chateau Slate	W
Lemonwood Yellow	Y

Interior trim codes
Fawn	709
Bright Blue	732
Red	747
Black	761
Turquoise	776
Bronze	787
Ivory	798

Options
13817 2 dr sport coupe, 8 cyl			$2,776.00
13867 Convertible, 8 cyl			2,962.00

Option number	Description	Quantity	Retail price*
A01	Soft Ray tinted glass (all windows)	77,021	$ 30.32/ 30.55
A02	Soft Ray tinted glass (windshield only)	190,741	19.34/ 19.50
A31	Power windows	3,164	99.33/ 100.10
A33	Power tailgate window	12,020	26.14/ 26.35
A39	Custom deluxe seatbelts w/front retractors	115,001	10.46/ 10.55
A41	4 way power front seat	938	69.01/ 69.55

Code	Description	Quantity	Price
A46	6 way power front seat	587	69.01/ 69.55
A51	Strato-Bucket seats	95,482	109.79/ 110.60
A81	Strato-Ease headrests	2,380	52.28/ 52.70
A82	Strato-Ease headrests	689	41.82/ 42.15
B90	Side window moldings	104	20.91/ 26.35
C06	Power-operated top	10,028	52.28/ 52.70
C08	Vinyl roof cover	32,507	73.19/ 73.75
C48	Heater & defroster deletion	7,861	70.17/ 70.70 CR
C50	Rear window defroster	5,089	20.91/ 21.10
C51	Rear window air deflector	1,517	18.82/ 19.00
C60	Four Season AC	60,814	353.41/ 356.00
D55	Console	82,477	47.05/ 47.40
F40	Special front & rear suspension	44,965	3.66/ 4.75
G66	Superlift rear shock absorbers	1,752	36.60/ 36.90
G76	3.36 ratio rear axle	1,265	2.09/ 2.15
G80	Positraction rear axle	67,905	36.60/ 36.90
G94	3.31 ratio rear axle	430	2.09/ 2.15
G96	3.55 ratio rear axle	2,577	2.09/ 2.15
H01	3.07 ratio rear axle	33	2.09/ 2.15
J50	Power brakes	33,695	41.82/ 42.15
J65	Metallic brakes	7,837	36.60/ 36.90
K02	Radiator fan—temp. controlled	3,365	15.68/ 15.80
K19	AIR equipment	54,166	44.44/ 44.75
K24	Closed positive engine ventilation	954	5.23/ 5.25
K66	Transistorized ignition system	2,013	73.19/ 73.75
K76	61 amp Delcotron generator	439	20.91/ 21.10

Code	Description	Quantity	Price
K79	42 amp Delcotron generator	867	10.46/ 10.55
K81	62 amp Delcotron generator	123	62.74/ 73.75
L26	140 hp Turbo-Thrift 230 6 cyl engine	66,943	26.14/ 26.35
L30	275 hp Turbo-Fire 327 V–8 engine	46,230	92.01/ 92.70
L34	360 hp Turbo-Jet 396 V–8 engine	24,811	104.56/ 105.35
L77	220 hp Turbo-Fire 283 V–8 engine	10,101	52.28/ 52.70
L78	375 hp Turbo-Jet 396 V–8 engine	3,099	235.26/ 237.00
M01	HD clutch	495	5.23/ 5.30
M10	Overdrive transmission	2,803	115.02/ 115.90
M20	4 speed transmission	73,022	104.56/ 184.35
M21	4 speed close-ratio transmission	5,012	104.56/ 105.35
M35	Powerglide transmission	292,437	192.92/ 194.85
M55	Transmission oil cooler	28	15.68/ 15.80
N10	Dual exhaust	11,102	20.91/ 21.10
N33	Comfortilt steering wheel	6,115	41.82/ 42.15
N34	Sports-styled steering wheel	7,537	31.37/ 31.60
N40	Power steering	185,274	83.65/ 73.75
N96	Mag-style wheel covers	10,814	73.19/ 73.75
P01	Wheel covers	223,842	20.91/ 21.10
P02	Simulated wire wheel covers	6,974	73.19/ 73.75
P19	Spare wheel lock	2,101	5.23/ 5.30
P58	7.35–14–4 pr whitewall tubeless tires	106,413	38.93/ 39.15
P61	7.75–14–4 pr nylon whitewall tubeless tires	4,897	72.19/ 72.40
P62	7.75–14–4 pr whitewall tubeless tires	63,795	53.39/ 53.60
P67	6.95–14–4 pr whitewall tubeless tires	76,821	28.03/ 28.20
T60	HD battery	20,503	7.32/ 7.40

U03	Tri-Volume horn	2,653	13.59/ 13.70
U14	Special instrumentation	32,436	31.37/ 79.00
U16	Tachometer	8,025	47.05/ 47.40
U63	Push-button radio	313,897	56.99/ 57.40
U69	Push-button AM/FM radio	5,689	132.79/ 133.80
U73	Manual rear antenna	66,846	9.41/ 9.50
U75	Power rear antenna	962	28.23/ 28.45
U80	Auxiliary speaker	54,750	70.60/ 147.00
V01	HD radiator	25,108	10.46/ 10.55
V31	Front bumper guard	16,928	9.41/ 9.50
V32	Rear bumper guard	13,067	9.41/ 9.50
V55	Roof luggage carrier	7,299	41.82/ 42.15
V74	Traffic hazard warning switch	35,398	11.50/ 11.60
Z19	Comfort & convenience equipment	18,487	20.91/ 26.35
720	Fawn vinyl interior trim	1,775	5.23/ 5.30
761	Black vinyl interior trim	95,102	10.46/ 10.55
950	Two-tone paint combination	18,060	15.68/ 15.80

*Prices reflect a six percent & a seven percent excise tax.

Facts

In 1966, the Chevelle SS 396 became a separate model line, identified by 13817 for the two-door coupe and 13867 for the convertible in the Chevelle's identification number.

The Chevelle got new front and rear styling treatments and a tunneled roofline. The Chevelle SS 396s got a blacked-out grille and rear taillight panel, simulated side-facing hood scoops, and sill and rear fender moldings. An SS 396 identification was used on the grille, Super Sport lettering appeared on both rear fenders and Chevelle SS 396 identification was displayed on the rear taillight panel. The full-size wheel covers were optional; the standard 14x6 in. steel rims were painted body color and came with the standard hubcaps and 7.75x14 Red Line tires.

In the interior, bench seats were standard equipment, bucket seats optional. The dash was redesigned.

The heart of the SS 396 was the 396 ci engine, which was available in three forms. The standard L35 396 ci engine came with the two-bolt main block, small oval-port cylinder heads, a Quadrajet four-barrel carburetor mounted on a cast-iron intake and a hydraulic camshaft, for a 325 hp rating. Standard transmission was a three-speed manual, with a four-speed manual and a Powerglide automatic optional.

The L34 396 ci was next on the list, rated at 360 hp. It came with a higher-performance camshaft, a four-bolt main cap cylinder block and a Holley carburetor.

The 375 hp L78 came with large rectangular-port cylinder heads that had larger (2.19 in. versus 2.06 in.) intake valves, an aluminum high-rise intake manifold with a Holley carburetor, a solid-lifter camshaft and a higher compression ratio (11.0:1 versus 10.25:1 for the other two 396s). The 375 hp 396 ci was available only with a four-speed manual transmission. Air conditioning was an option that could be ordered in conjunction with the L78.

All 396 engines came with chrome valve covers, air cleaner and oil filler cap.

The 1966 Chevelle Malibu SS 396 Coupe.

Chapter 13

1967 Chevelle Malibu SS 396

Production

By engine
2 dr coupe & convertible,
L35 396 ci V-8   45,218
2 dr coupe convertible,
L34 396 ci V-8   17,176
2 dr coupe & convertible,
L78 396 ci V-8   612
Total   63,006

By body style
13817 2 dr coupe   59,685
13867 2 dr convertible   3,321
Total   63,006

Serial numbers

Description
138177A100001
13817 — Model number (13817-2 dr coupe, 13867-2 dr convertible)
7 — Last digit of model year (1967)
A — Assembly plant (A-Atlanta, B-Baltimore, G-Framingham,
K-Kansas City, Z-Fremont)
100001 — Consecutive sequence number

Location
On plate attached to left front door hinge post.

Engine and transmission suffix codes
ED — 396 ci V-8 4 bbl 325 hp, 3 or 4 speed manual
EH — 396 ci V-8 4 bbl 325 hp, 3 or 4 speed manual w/AIR
EK — 396 ci V-8 4 bbl 325 hp, Powerglide automatic
EM — 396 ci V-8 4 bbl 325 hp, Powerglide automatic w/AIR
ET — 396 ci V-8 4 bbl 325 hp, Turbo Hydra-matic automatic
EV — 396 ci V-8 4 bbl 325 hp, Turbo Hydra-matic automatic w/AIR
EF — 396 ci V-8 4 bbl 350 hp, 3 or 4 speed manual
EJ — 396 ci V-8 4 bbl 350 hp, 3 or 4 speed manual w/AIR
EL — 396 ci V-8 4 bbl 350 hp, Powerglide automatic
EN — 396 ci V-8 4 bbl 350 hp, Powerglide automatic w/AIR
EU — 396 ci V-8 4 bbl 350 hp, Turbo Hydra-matic automatic
EW — 396 ci V-8 4 bbl 350 hp, Turbo Hydra-matic automatic w/AIR
EG — 396 ci V-8 4 bbl 375 hp, 4 speed manual
EX — 396 ci V-8 4 bbl 375 hp, 4 speed manual w/AIR

Carburetors
396 ci 325 hp — 7027201
396 ci 325 hp w/automatic — 7027200
396 ci 350 hp — 3908957 (Holley R-3837A)

396 ci 350 hp w/AIR — 3908959 (Holley R-3839A)
396 ci 350 hp w/automatic — 3908956 (Holley R-3836A)
396 ci 350 hp w/AIR & automatic — 3908957 (Holley R-3838A)
396 ci 375 hp — 3916143 (Holley R-3910A)
396 ci 375 hp w/AIR — 3916145 (Holley R-3911A)

Distributors
396 ci — 1111169,
396 ci 350 hp — 1111170
396 ci 375 hp — 1111170

Cylinder head casting numbers
396 ci 325/350 hp — 3909802
396 ci 375 hp — 3904391

Exterior color codes
Tuxedo Black	AA
Ermine White	CC
Nantucket Blue	DD
Deepwater Blue	EE
Marina Blue	FF
Granada Gold	GG
Mountain Green	HH
Emerald Turquoise	KK
Tahoe Turquoise	LL
Royal Plum	MM
Madeira Maroon	NN
Bolero Red	RR
Sierra Fawn	SS
Capri Cream	TT
Butternut Yellow	YY

Interior trim codes
Bright Blue	723/731*
Blue	729/738*
Red	747/750*
Black	761/763*
Turquoise	776/778*
Gold	783/784*

*Bucket seats.

Convertible top color codes
White	AA
Black	BB
Medium Blue	—

Vinyl top color codes
Black	BB
Light Fawn	—

Options*
13817 2 dr sport coupe	$2,825.00
13867 2 dr convertible	3,033.00

Option number	Description	Quantity	Retail price
AL5	Deluxe rear center seatbelt	2,901	$ 7.90
AS1	Std shoulder harness	411	23.20
A01	Soft Ray tinted glass (all windows)	80,859	30.55
A02	Soft Ray tinted glass (windshield only)	177,059	21.10
A31	Power windows	2,886	100.10
A33	Power rear window (wagons)	12,954	31.60
A41	4 way front seat control	889	69.55
A49	Custom deluxe seatbelts	9,182	6.35
A51	Strato-Bucket seats	79,475	110.60
A81	Headrests (w/Strato-Bucket front seats)	1,609	52.70
A82	Headrests (w/std bench front seat)	924	42.15

A85	Deluxe shoulder harness	1,064	26.35
B37	2 front & 2 rear color-keyed floor mats	37,421	10.55
B55	Extra-thick foam front seat cushion	2,334	7.40
B90	Side window moldings	1,511	21.10-26.35
B93	Door edge guards	65,316	3.20-6.35
C06	Power tops	7,642	52.70
C08	Vinyl roof cover	70,294	73.75
C48	Heater & defroster deletion	6,318	70.70 CR
C50	Rear window defroster	7,417	21.10
C60	Four Season AC	75,063	356.00
D33	Remote control outside LH mirror	9,760	9.50
D55	Front compartment console	71,482	47.40
F40	Special front & rear suspension	39,321	4.75
G66	Superlift rear shock absorbers	1,885	36.90
G75	3.70 ratio rear axle equipment	803	2.15
G76	3.36 ratio rear axle	1,191	2.15
G80	Positraction rear axle	57,988	42.15
G92	3.08 ratio rear axle	79	2.15
G94	3.31 ratio rear axle	1,209	2.15
G96	3.55 ratio rear axle	2,917	2.15
G97	2.73 ratio rear axle	51	2.15
J50	Power brakes	36,992	42.15
J52	Front disc brakes	5,153	79.00
J65	Brakes w/metallic facings	1,711	36.90
K02	Temperature-controlled radiator fan	2,564	15.80
K19	AIR equipment	42,147	44.75
K24	Closed positive engine ventilation	40,515	5.25
K76	61 amp Delcotron generator	532	21.10
K79	42 amp Delcotron generator	835	10.55
L22	155 hp Turbo-Thrift 250 engine	41,778	26.35
L30	275 hp Turbo-Fire 327 engine	43,440	92.70
L78	350 hp Turbo-Jet 396 engine	612	105.35
L79	325 hp Turbo-Fire 327 engine	4,048	198.05
M01	HD clutch	1,925	5.30-10.55

M10	Overdrive transmission	1,878	115.90
M13	Special fully synchronized 3 speed transmission	2,755	79.00
M20	4 speed wide-range transmission	48,354	105.35– 184.35
M21	4 speed close-ratio transmission	12,886	105.35– 184.35
M22	4 speed transmission	—	—
M35	Powerglide transmission	267,510	115.90– 194.85
M40	Turbo Hydra-matic	13,272	147.45
N10	Dual exhaust	7,365	21.10
N30	Deluxe steering wheel	2,387	4.25– 7.40
N33	Comfortilt steering wheel	6,586	42.15
N34	Sports-styled walnut-grained steering wheel w/plastic rim	5,462	31.60
N40	Power steering	201,865	84.30
PW7	F70-14-4 pr white stripe SS tubeless tires	22,678	NC
PW8	F70-14-4 pr red stripe SS tubeless tires	2,631	Std
P01	Brightmetal wheel covers	229,684	21.10
P02	Simulated wire wheel covers	5,484	73.75
P58	7.35-14-4 pr whitewall tubeless tires	160,092	31.35
P61	7.75-14-4 pr whitewall tubeless tires	1,027	45.80
P62	7.75-14-4 pr whitewall tubeless tires	67,771	31.35
P67	6.65-14-8 pr whitewall tubeless tires	5	—
T15	7.75-14-8 pr whitewall tubeless tires	287	79.20
T60	HD battery	19,298	7.40
U03	Tri-Volume horn	1,191	13.70
U15	Speed warning indicator	5,945	10.55
U16	Tachometer	3,653	47.40
U25	Luggage compartment lamp	23,498	2.65
U26	Underhood lamp	25,478	2.65
U27	Glove compartment lamp	1,871	2.65
U28	Ashtray lamp	28,582	1.60
U29	Instrument panel courtesy lights	28,252	4.25
U35	Electric clock	1,038	15.80
U57	Stereo tape system	5,194	128.50
U63	Push-button control radio	311,388	57.40– 70.60

U69	Push-button control		133.80–
	AM/FM radio	5,915	147.00
U73	Rear antenna	46,911	9.50
U80	Auxiliary speaker	57,054	13.20
V01	HD radiator	12,695	10.55
V31	Front bumper guard	24,655	12.65
V32	Rear bumper guard	17,603	12.65
V55	Luggage carrier equipment		
	(wagons)	8,834	42.15
Z29	Special bodyside accent		
	stripes	49,580	—
761	All-vinyl interior trim	142,000	10.55
794	Deluxe cloth interior trim	1,648	68.50
950	Two-tone exterior paint		
	combinations	13,901	15.80
—	Appearance Guard Group		29.55–
			48.55
—	Auxiliary Lighting Group		6.90–
			13.80
—	Foundation Group		80.60
—	Operating Convenience		
	Group		30.60
—	Station Wagon Convenience		
	Group		92.75

*Options are for all Chevelle models as of September 29, 1966.

Facts

The 1967 Chevelle SS 396 was mildly facelifted using different grille and rear end treatments. The SS identification and features were basically the same, the only differences being the addition of wheelwell moldings and side accent stripes. The rear fender Super Sport emblem had *Super* on the top and *Sport* on the bottom.

The standard Red Line tires were lower-profile F70x14s.

In the interior, SS identification was found on the steering wheel and on the dash.

Front disc brakes were optional for the first time, but they required the use of the optional rally wheels.

Other improvements included government-mandated dual master cylinder brakes and the availability of the new three-speed Turbo Hydra-matic automatic transmission on all engines. Other transmission choices remained unchanged from those offered in 1966.

Engine availability was the same as in 1966, but the L34 396 ci engine was downgraded by 10 hp, now rated at 350 hp.

The 1967 Chevelle Malibu SS 396 Coupe.

Chapter 14

1968 Chevelle SS 396

Production

By engine
2 dr coupe, convertible
& pickup, L35 396 ci   45,553
2 dr coupe , convertible
& pickup, L34 396 ci   12,481
2 dr coupe, convertible
& pickup, L78 396 ci   4,751
Total   62,785

By body style
13837 2 dr coupe   55,309
13867 2 dr convertible   2,286
13880 2 dr pickup
  delivery   5,190
  Total   62,785

Serial numbers

Description
138378A100001
13837 — Model number (13837-2 dr coupe, 13867-2 dr convertible, 13880-2 dr passenger pickup)
8 — Last digit of model year (1968)
A — Assembly plant (A-Atlanta, B-Baltimore, G-Framingham, K-Kansas City, Z-Fremont)
100001 — Consecutive sequence number

Location
On plate attached to driver's side of dash, visible through the windshield.

Engine and transmission suffix codes
ED — 396 ci V-8 4 bbl 325 hp, 3 or 4 speed manual
EK — 396 ci V-8 4 bbl 325 hp, Powerglide automatic
ET — 396 ci V-8 4 bbl 325 hp, Turbo Hydra-matic automatic
EF — 396 ci V-8 4 bbl 350 hp, 3 or 4 speed manual
EL — 396 ci V-8 4 bbl 350 hp, Powerglide automatic
EU — 396 ci V-8 4 bbl 350 hp, Turbo Hydra-matic automatic
EG — 396 ci V-8 4 bbl 375 hp, 4 speed manual

Carburetors
396 ci 325 hp — 7028211
396 ci 325 hp w/automatic — 7028210
396 ci 350 hp — 7028217
396 ci 350 hp w/automatic — 7028218
396 ci 375 hp — 3923289 (Holley R-4053A)

Distributors
396 ci 325 hp — 1111169
396 ci 350 hp — 1111145

396 ci 350 hp w/automatic — 1111169
396 ci 375 hp — 1111170

Cylinder head casting numbers

396 ci 325/350 hp — 3917215
396 ci 375 hp — 3919840

Exterior color codes

Tuxedo Black	AA
Ermine White	CC
Grotto Blue	DD
Fathom Blue	EE
Island Teal	FF
Ash Gold	GG
Grecian Green	HH
Tripoli Turquoise	KK
Teal Blue	LL
Cordovan Maroon	NN
Seafrost Green	PP
Matador Red	RR
Palomino Ivory	TT
Sequoia Green	VV
Butternut Yellow	YY

Interior trim codes

Black	765/766*
Gold	754/756*
Red	795
Teal	755/757*
Parchment/Black	793/794*
*Bucket seats.	

Convertible top color codes

White	AA
Black	BB
Blue	—

Vinyl top color codes

Black	BB
White	AA

Options*

13837 Sport coupe	$2,875.00
13867 Convertible	3,102.00

Option number	Description	Quantity	Retail price
AS1	2 front seatbelts & shoulder belts	18	$ 23.20
AS4	2 front & 2 rear custom deluxe shoulder belts	130	52.70
AS5	2 front & 2 rear std-style shoulder belts	44	46.40
A01	Soft Ray tinted glass (all windows)	163,309	34.80
A02	Soft Ray tinted glass (windshield only)	120,409	23.20
A31	Power windows	3,595	100.10
A33	Power tailgate window	17,348	31.60
A39	Custom deluxe seatbelts	1,302	7.90–9.50
A51	Strato-Bucket seats	71,587	110.60
A81	Head restraints (w/Strato-Bucket front seats)	2,426	52.70
A82	Head restraints (w/full-width front seat)	1,571	42.15
A85	Custom deluxe shoulder belts	122	26.35
B37	Color-keyed floor mats	67,199	10.55
B55	Extra-thick foam seat cushion	2,783	7.40

Code	Description	Qty	Price
B90	Side window moldings	6,436	21.10–31.60
B93	Door edge guards	103,407	4.25–7.40
C06	Power top (convertible only)	6,448	52.70
C08	Vinyl roof cover	142,220	84.30
C24	Hide-a-Way windshield wipers	688	19.00
C50	Rear window defroster	8,301	21.10
C51	Rear window air deflector	3,440	19.00
C60	Four Season AC	114,066	360.20
D33	Remote control outside rearview mirror	10,266	9.50
D55	Console	65,338	50.60
D96	Accent striping	18,934	29.50
F40	Special front & rear suspension	28,747	4.75
G66	Superlift air-adjustable shock absorbers	1,973	42.15
G75	3.70 ratio rear axle	280	2.15
G76	3.36 ratio rear axle	1,419	2.15
G80	Positraction rear axle	60,283	42.15
G92	3.08 ratio rear axle	—	2.15
G94	3.31 ratio rear axle	1,343	2.15
G96	3.55 ratio rear axle	3,697	2.15
G97	2.73 ratio rear axle	2,192	2.15
H01	3.07 ratio rear axle	534	2.15
H05	3.73 ratio rear axle	5,553	2.15
J50	Power brakes	59,937	42.15
J52	Power disc brakes	12,835	100.10
KD5	HD engine ventilation	36	6.35
K02	Temperature-controlled fan	1,823	15.80
K30	Cruise-Master speed control	743	52.70
K76	61 amp Delcotron generator	556	5.30–26.35
K79	42 amp Delcotron generator	4,338	10.55
L22	155 hp Turbo-Thrift 250 6 cyl engine	43,083	26.35
L30	275 hp Turbo-Fire 327 V-8 engine	33,890	92.70
L34	350 hp Turbo-Jet 396 V-8 engine (SS 396 only)	12,481	105.35
L73	250 hp Turbo-Fire 327 V-8 engine	36,035	63.20
L78	375 hp Turbo-Jet 396 V-8 engine	4,751	237.00
L79	325 hp Turbo-Fire 327 V-8 engine	4,082	198.05

Code	Description	Qty	Price
M01	HD clutch	2,291	5.30–10.55
M10	Overdrive transmission	1,500	115.90
M13	3 speed transmission	4,027	79.00
M20	4 speed wide-range transmission	38,933	184.35
M21	4 speed close-ratio transmission	11,208	184.35
M22	HD 4 speed close-ratio transmission (SS 396 only)	1,049	237.00
M35	Powerglide transmission	334,061	184.35–194.85
M40	Turbo Hydra-matic transmission (SS 396 models only)	21,539	237.00
N10	Dual exhaust	8,476	27.40
N30	Deluxe steering wheel	6,435	4.25–7.40
N33	Comfortilt steering wheel	7,450	42.15
N34	Sports-styled steering wheel	4,698	31.60
N40	Power steering	289,656	94.80
N95	Simulated wire wheel covers	4,598	73.75
N96	Mag-style wheel covers	6,063	73.75
PA2	Mag-spoke wheel covers	999	73.75
PN5	7.75-14-8 pr whitewall tubeless tires	594	73.00
PW7	F70-14-4 pr white stripe tubeless tires	28,685	51.00
PW8	F70-14-4 pr red stripe tubeless tires	5,377	51.00
P01	Wheel covers	266,182	21.80
P58	7.35-14-4 pr whitewall tubeless tires	195,210	31.35
P62	7.75-14-4 pr whitewall tubeless tires	77,601	46.00
T60	HD battery	25,434	7.40
U03	Tri-Volume horn	1,125	13.70
U14	Special instrumentation	19,393	94.80
U15	Speed warning indicator	4,546	10.55
U25	Luggage compartment lamp	24,657	6.85–13.70
U26	Underhood lamp	29,802	6.85–13.70
U27	Glove compartment lamp	2,099	6.85–13.70
U28	Ashtray lamp	29,658	6.85–13.70
U29	Instrument panel courtesy lights	28,482	6.85–13.70

U35	Electric clock	74,057	15.80
U46	Light monitoring system	2,757	26.35
U57	Stereo tape system	5,983	133.80
U63	Push-button radio	370,223	61.10
U69	Push-button AM/FM radio	15,964	133.80
U73	Manual rear antenna	35,218	9.50
U79	AM/FM stereo radio	1,707	239.15
U80	Auxiliary speaker	52,949	13.20
V01	HD radiator	10,574	13.70
V31	Front bumper guards	34,166	15.80
V32	Rear bumper guards	25,786	15.80
V55	Luggage carrier (wagons)	12,289	44.25
ZJ7	Rally wheels	31,149	31.60
ZJ9	Auxiliary lighting	30,036	6.85–13.70
780	Deluxe cloth interior trim	1,664	—
793	Vinyl interior trim	177,593	10.55
950	Two-tone paint combinations	16,601	—

*Options are for all Chevelle models as of September 1967.

Facts

The Chevelle received a major restyle in 1968, giving the car more of a fastback look. The Chevelle SS 396 continued to be a separate, distinct model line identified by the model numbers 13837, 13867 and 13880. For 1968, the El Camino passenger pickup was included as an SS 396 model.

Super Sport features were as follows: blacked-out grille with SS 396 emblem, blacked-out taillight panel with SS 396 emblem, black lower body area on light-colored cars, domed hood and lower rear fender moldings. The numerals 396 were incorporated on the front-fender-mounted signal lamps. Early cars included the letters SS with the numerals 396. Full-size Super Sport wheel covers were optional. In the interior, an SS emblem was located on the dash panel above the glovebox door.

Engine choice was unchanged from that in 1967, limited to the three 396 ci engines.

Beginning with 1968, the VIN plate was located on the dash and was visible through the windshield. Front and rear side marker lights were used on all 1968 Chevelles.

Hidden windshield wipers were standard on the SS 396.

The 1968 Chevelle 396s got improved finned front brake drums. The optional front disc brakes were available in conjunction with the rally wheels.

From 1968 on, all engines used a spin-on oil filter rather than the previous canister setup.

The 1968 Chevelle SS 396 Sport Coupe.

Chapter 15

1969 Chevelle SS 396

Production

2 dr coupe & convertible,
L35 396/402 ci V-8   59,786
2 dr coupe & convertible,
L34 396/402 ci V-8   17,358
2 dr coupe & convertible,
L78 396/402 ci V-8   9,486*

2 dr coupe & convertible,
L89 396/402 ci V-8   400
Total   87,030
  *Includes approximately 323 of the L72 427 ci COPO Chevelles.

Serial numbers

Description
136379A300001
13637 — Model number (13427-2 dr convertible, 13437-2 dr coupe, 13637-2 dr coupe, 13667-2 dr convertible)
9 — Last digit of model year (1969)
A — Assembly plant (A-Atlanta, B-Baltimore, G-Framingham, K-Kansas City, Z-Fremont)
300001 — Consecutive sequence number

Location
On plate attached to driver's side of dash, visible through the windshield.

Engine and transmission suffix codes
JA — 396 ci V-8 4 bbl 325 hp, 3 or 4 speed manual
CJA — 402 ci V-8 4 bbl 325 hp, 3 or 4 speed manual
JK — 396 ci V-8 4 bbl 325 hp, Turbo Hydra-matic automatic
CJK — 402 ci V-8 4 bbl 325 hp, Turbo Hydra-matic automatic
JC — 396 ci V-8 4 bbl 350 hp, 3 or 4 speed manual
CJC — 402 ci V-8 4 bbl 350 hp, 3 or 4 speed manual
JE — 396 ci V-8 4 bbl 350 hp, Turbo Hydra-matic automatic
CJE — 402 ci V-8 4 bbl 350 hp, Turbo Hydra-matic automatic
JD — 396 ci V-8 4 bbl 375 hp, 4 speed manual
CJD — 402 ci V-8 4 bbl 375 hp, 4 speed manual
KF — 396 ci V-8 4 bbl 375 hp, Turbo Hydra-matic automatic
CKF — 402 ci V-8 4 bbl 375 hp, Turbo Hydra-matic automatic
KG — 396 ci V-8 4 bbl 375 hp, 4 speed manual w/L89
CKG — 402 ci V-8 4 bbl 375 hp, 4 speed manual w/L89
KH — 396 ci V-8 4 bbl 375 hp, Turbo Hydra-matic automatic w/L89
CKH — 402 ci V-8 4 bbl 375 hp, Turbo Hydra-matic automatic w/L89

Carburetors
396 ci 325 hp — 7029215
396 ci 325 hp w/automatic — 7029204
396 ci 350 hp — 7029215

396 ci 350 hp w/automatic — 7029204
396 ci 375 hp — 3959164 (Holley R-4346A)

Distributors
396 ci — 1111497
396 ci 350 hp — 1111499
396 ci 375 hp — 1111499

Cylinder head casting numbers
396 ci 325/350 hp — 3931063
396 ci 375 hp — 3919840,
  3919842 aluminum

Exterior color codes

Color	Code
Tuxedo Black	10
Butternut Yellow	40
Dover White	50
Dusk Blue	51
Garnet Red	52
Glacier Blue	53
Azure Turquoise	55
Fathom Green	57
Frost Green	59
Burnished Brown	61
Champagne	63
Olympic Gold	65
Burgundy	67
Cortez Silver	69
LeMans Blue	71
Monaco Orange	72
Daytona Yellow	76

Interior trim codes

Color	Models 13427 & 13437	Model 13637	Model 13667
Black	751/752*	753/755*/756**	—
Blue	760	—	764*
Green	786	784*/785**	—
Dark Green	—	782	—
Red	—	787*/788**	788**
Turquoise	779	—	—
Parchment/Black	—	790*	791**

*Vinyl bench seats.
**Vinyl bucket seats.

Convertible top color codes

White	AA
Black	BB

Vinyl top color codes

Black	BB
Parchment	EE
Dark Blue	CC
Dark Brown	FF
Midnight Green	GG

Options

13637 Sport coupe	$2,673.00
13667 Convertible	2,872.00

Option number	Description	Quantity	Retail price
AR1	Less head restraint	774	—
AS1	Std shoulder harness	45	$ 23.20
AS4	Deluxe rear seat shoulder harness	91	26.35
AS5	Std rear seat shoulder harness	206	23.20

A01	Tinted glass (all windows)	300,035	36.90
A02	Tinted glass (windshield)	4,470	
A31	Electric control windows	4,583	105.35
A33	Electric tailgate window	17,542	34.80
A39	Custom deluxe front & rear seatbelts	789	9.00–10.55
A51	Strato-type front bucket seat	79,871	121.15
A85	Deluxe shoulder harness	168	26.35
A93	Vacuum-operated door locks	1,206	44.80–68.50
BX4	Bodyside molding	11,377	26.35
B37	Floor mats	84,085	11.60
B90	Door & window frame molding	7,633	21.90–26.35
B93	Door edge guards	133,295	4.25–7.40
CE1	Headlamp washer	562	15.80
C06	Electric control folding top	6,211	52.70
C08	Exterior soft trim roof cover	201,671	89.55
C24	Special windshield wiper	706	19.00
C50	Rear window defroster	15,372	22.15–32.65
C51	Rear window air deflector	5,078	—
C60	Deluxe AC	168,449	376.00
D33	Remote control outside mirror	20,025	10.55
D34	Vanity Visor mirror	18,576	3.20
D55	Front compartment floor console	72,201	53.75
D96	Wide side paint stripe	37,280	26.35
F40	Special front & rear suspension	28,749	5.30–16.90
F41	Special performance front & rear suspension	722	29.50
GT1	2.56 ratio rear axle	784	2.15
G76	3.36 ratio rear axle	2,551	2.15
G80	Positraction rear axle	73,397	42.15
G82	4.56 ratio rear axle	1	2.15
G84	4.10 ratio rear axle	6,250	2.15
G92	3.08 ratio rear axle	1,441	2.15
G94	3.31 ratio rear axle	1,617	2.15
G96	3.55 ratio rear axle	4,369	2.15
G97	2.73 ratio rear axle	1,268	2.15
H01	3.07 ratio rear axle	2,369	2.15
H05	3.73 ratio rear axle	8,588	2.15
J50	Vacuum power brake equipment	164,075	42.15
J52	Power front disc brakes	110,651	64.25
KD5	HD closed positive ventilation	129	6.35

Code	Description	Production	Price
K02	Fan drive equipment	1,822	15.80
K05	Engine block heater	3,449	10.55
K79	42 amp AC generator	1,605	10.55
K85	63 amp AC generator	1,080	5.30–26.35
LM1	350 ci V-8 engine (regular fuel)	23,315	68.50
L22	250 ci L-6 engine	28,932	26.35
L34	396 ci Hi-Performance V-8 engine	17,358	121.15
L35	396 ci V-8 engine	59,786	
L48	350 ci V-8 engine	30,099	68.50
L65	350 ci 2 bbl V-8 engine	53,969	21.10
L78	396 ci Special Hi-Performance V-8 engine	9,486	252.80
L89	Aluminum cylinder heads	400	647.75
MC1	HD 3 speed transmission	15,748	79.00
M20	4 speed transmission	44,950	184.80
M21	4 speed close-ratio transmission	13,786	184.80
M22	HD 4 speed transmission (375 hp only)	1,276	264.00
M35	Powerglide transmission	226,784	163.70–174.25
M38	300 Deluxe 3 speed automatic transmission	135,503	200.65
M40	3 speed automatic transmission	32,031	190.10–221.80
NC8	Chambered pipe exhaust system	4,143	15.80
N10	Dual-exhaust equipment	13,358	30.55
N33	Tilt-type steering wheel	10,718	45.30
N34	Wood-grain plastic steering wheel	7,515	34.80
N40	Power steering	366,017	100.10–105.35
N95	Simulated wire wheel trim cover	1,909	73.75
N96	Simulated magnesium wheel trim cover type A	2,121	73.75
PA2	Simulated magnesium wheel trim cover type B	947	73.75
PK2	G78-14-4 pr B/B whitewall tires	1,441	—
PK4	G70-14-4 pr red stripe tires	4,436	—
PL5	F70-14-4 pr white letter tires	60,513	57.35–72.15
PN5	7.75-14-8 pr whitewall tires	334	—
PQ7	8.25-14-4 pr whitewall tires	168	—

Code	Description	Quantity	Price
PR3	8.25-14-8 pr whitewall tires	4,264	—
PW7	F70-14-4 pr white stripe tires	19,799	57.55-72.30
PW8	F70-14-4 pr red stripe tires	6,243	57.55-72.30
PX6	F78-14-4 pr B/B whitewall tires	896	72.05-86.80
PX8	F78-14-4 pr whitewall tires	613	—
PX9	G70-14-4 pr white stripe tires	1,299	—
PY4	F70-14-4 pr B/B white stripe tires	5,127	26.25-98.35
PY5	F70-14-4 pr B/B red stripe tires	1,518	26.25-98.35
PY7	G70-14-4 pr B/B red stripe tires	257	—
P01	Wheel trim cover	243,191	21.10
P06	Wheel trim ring	1,055	21.10
P58	7.35-14-4 pr whitewall tires	173,217	33.45
P62	7.75-14-4 pr whitewall tires	61,331	48.20
P77	8.25-14-4 pr whitewall tires	20,258	
T60	HD battery	26,795	8.45-15.80
UF1	Map lamp	34,113	—
U05	Dual horn	1,471	—
U14	Instrument panel gauges	24,852	94.80
U15	Speed warning indicator	4,372	11.60
U25	Luggage compartment lamp	28,100	12.15-19.00
U26	Underhood lamp	34,114	12.15-19.00
U27	Glove compartment lamp	1,875	12.15-19.00
U28	Ashtray lamp	34,114	12.15-19.00
U29	Instrument panel courtesy lights	32,901	12.15-19.00
U35	Electric clock	75,277	15.80
U46	Lamp monitoring	2,041	26.35
U57	Tape player	12,635	133.80
U63	Push-button radio	428,294	61.10
U69	Push-button AM/FM radio	15,084	133.80
U73	Manual rear antenna	33,117	9.50
U79	Stereo equipment	3,247	239.10
U80	Auxiliary speaker	64,165	13.20
V01	HD radiator	8,588	14.75

V31	Front bumper guard equipment	27,003	15.80
V32	Rear bumper guard equipment	19,543	15.80
V55	Luggage carrier equipment	12,983	—
V75	Traction compound & dispenser	278	23.20
ZJ6	Special sport sedan	2,022	131.65
ZJ7	Special wheel, hubcap & trim ring	40,238	35.85
ZJ9	Auxiliary Lighting Group	34,116	12.15–19.00
ZK3	Deluxe seatbelts & front seat shoulder harness	48,792	12.15–13.70
Z25	SS equipment	86,307	347.60
796	Vinyl-coated trim	220,594	12.65
959	Two-tone color combination	21,690	23.20

Facts

Only a minor restyle occurred on the 1969 Chevelle to differentiate it from the all-new 1968 model. The front grille was restyled using a horizontal grille bar in the center with the bumper incorporating an opening in which the turn signal lamps were located. The rear taillights were larger and were angled inward.

The Chevelle SS 396 was no longer a separate model series. It became an option, RPO Z25, which was available only on the following Chevelle models:

13427 2 dr Chevelle 300 Deluxe Series coupe
13437 2 dr Chevelle 300 Deluxe Series coupe
13537 2 dr Malibu Series coupe
13667 2 dr convertible

The VIN of an original Chevelle SS 396 will have one of the above numbers as its first five digits.

Super Sport cars still came with the blacked-out grille and rear taillight panel. The wheelwells got bright moldings and, for the first time, chrome five-spoke Magnum wheels with SS center caps were standard. The grille, rear panel and front fenders carried SS 396 emblems.

In the interior, SS emblems were on the steering wheel and dash.

All 396 Chevelles were equipped with power front disc brakes as standard equipment, and if power steering was ordered, variable quick-ratio steering was included.

Engine availability was unchanged from that in 1968: the standard L35 325 hp 396 ci, the L34 350 hp 396 ci and the L78 375 hp 396 ci. During the model year, the 396's bore was increased from 4.094 in. to 4.125 in., resulting in 402 ci. However, the engine was still marketed as the SS 396 in 1969 and in subsequent years.

The Powerglide two-speed automatic was not available, having been supplanted by the Turbo Hydra-matic three-speed automatic.

The 1969 Chevelle SS 396 Coupe.

Chapter 16

1970 Chevelle SS 396 and SS 454

Production

2 dr coupe & convertible,
  L34 402 ci V-8   51,437
2 dr coupe & convertible,
  L78 402 ci V-8   2,144
2 dr coupe & convertible,
  L89 402 ci V-8   18

2 dr coupe & convertible,
  LS5 454 ci V-8   4,298
2 dr coupe & convertible,
  LS6 454 ci V-8   4,475
  Total   62,372

Serial numbers

Description
136370A100001
13637— Model number (13637-2 dr coupe, 13667-2 dr convertible)
0 — Last digit of model year (1970)
A — Assembly plant (A-Atlanta, B-Baltimore, F-Flint, K-Leeds City, L-Van Nuys, 1-Oshawa)
100001 — Consecutive sequence number

Location
On plate attached to driver's side of dash, visible through the windshield.

Engine and transmission suffix codes
CTX — 402 ci V-8 4 bbl 350 hp, 4 speed manual
CTW — 402 ci V-8 4 bbl 350 hp, Turbo Hydra-matic automatic
CKO — 402 ci V-8 4 bbl 375 hp, 4 speed manual
CTY — 402 ci V-8 4 bbl 375 hp, Turbo Hydra-matic automatic
CKT — 402 ci V-8 4 bbl 375 hp, 4 speed manual w/aluminum heads
CKP — 402 ci V-8 4 bbl 375 hp, Turbo Hydra-matic automatic w/aluminum heads
CRT — 454 ci V-8 4 bbl 360 hp, 4 speed manual
CRQ — 454 ci V-8 4 bbl 360 hp, Turbo Hydra-matic automatic
CRV — 454 ci V-8 4 bbl 450 hp, 4 speed manual
CRR — 454 ci V-8 4 bbl 450 hp, Turbo Hydra-matic automatic

Carburetors
402 ci — 7047000
402 ci 375 hp w/EEC — 3967477 (Holley R-4557A)
402 ci 375 hp w/EEC — 3967479 (Holley R-4491A)
402 ci 375 hp w/EEC & automatic — 3969894 (Holley R-4556A)
402 ci 375 hp w/EEC & automatic — 3969898 (Holley R-4492A)
454 ci 360 hp — 7040501
454 ci 360 hp w/automatic — 7040500

454 ci 450 hp — 3967477 (Holley R-4803A)
454 ci 450 hp w/automatic — 3969898 (Holley R-4802A)

Distributors
402 ci — 11112000
402 ci 375 hp — 1112000
454 ci 360 hp — 1108418
454 ci 360 hp w/automatic —
1108430
454 ci 450 hp — 1111437

Cylinder head casting numbers
402 ci 350 hp — 3964290
402 ci 375 hp —3964291,
3946074 aluminum
454 ci 360 hp — 3964290
454 ci 450 hp — 3964291

Exterior color codes

Color	Code
Classic White	10
Cortez Silver	14
Shadow Gray	17
Tuxedo Black	19
Astro Blue	25
Fathom Blue	28
Misty Turquoise	34
Green Mist	45
Forest Green	48
Gobi Beige	50
Champagne Gold	55
Autumn Gold	58
Desert Sand	63
Cranberry Red	75
Black Cherry	78

Interior trim codes

Color	Coupe	Convertible
Black	753/755*	755*/756**
Medium Blue	762	764*/765**
Saddle	770*/771**	770*/771**
Dark Green	782	—
Medium Gold	776	—
Ivory	790*/791**	790*/791**
Red	—	788**

*Vinyl bench seats.
**Vinyl bucket seats.

Convertible top color codes
White   AA
Black   BB

Vinyl top color codes
White   AA
Black   BB
Dark Blue   CC
Dark Green   GG
Dark Gold   HH

Options
13537 Sport coupe $2,719.00
13567 Convertible   2,972.00

Option number	Description	Quantity*	Retail price
AK1	Deluxe seatbelts & front shoulder harness	78,206	$ 12.15–13.70
AQ2	Electric seatback lock release	663	
AS4	Deluxe rear seat shoulder harness	273	26.35

Code	Description	Quantity	Price
AU3	Electric door locks	5,482	44.80-68.50
A01	Tinted glass (all windows)	416,493	36.90
A02	Tinted glass (windshield)	10,994	—
A31	Electric control windows	20,792	105.35
A33	Electric tailgate window	17,621	
A39	Custom deluxe front & rear seatbelts	754	9.00-10.55
A41	4 way electric control front seat	637	
A46	4 way electric control front bucket seat	546	
A51	Strato-type front bucket seat	99,138	121.15
A85	Deluxe shoulder harness	122	26.35
A90	Electric control rear compartment lid release	987	
B37	Floor mats	140,880	11.60
B85	Belt reveal molding	906	
B90	Door & window frame molding (4 dr)	5,203	26.35
B93	Door edge guards	197,408	4.25-7.40
CD2	Windshield washer fluid level	40,630	—
CD3	Electro Tip windshield wiper	331	19.00
C06	Electric control folding top	6,626	—
C08	Exterior soft trim roof cover	287,052	94.80
C50	Rear window defroster	39,953	26.35-36.90
C51	Rear window air deflector	7,713	
C60	Deluxe AC	303,326	376.00
D33	Remote control outside mirror	68,830	10.55
D34	Vanity Visor mirror	37,287	3.20
D55	Front compartment floor console	90,436	53.75
D88	Sport stripe	39,539	68.50
F40	Special front & rear suspension	17,255	16.90
F41	Special performance front & rear suspension	56,051	29.50
G67	Rear shock absorber level control	3,929	—
G80	Positraction rear axle	64,169	42.15
JL2	Front disc brakes	139,856	64.25
J50	Vacuum power brake equipment	180,113	42.15
K05	Engine block heater	6,068	10.55
K30	Speed & cruise control	730	57.95

K85	63 amp AC generator	2,447	26.35
LF6	400 ci 2 bbl V-8 engine	16,595	50.00
LS3	400 ci V-8 engine	9,338	162.20
LS6	454 ci Special Hi-Performance V-8 engine	4,475	263.30
L48	350 ci V-8 engine	85,212	68.50
L65	350 ci 2 bbl V-8 engine	112,686	21.10
L78	396 ci Special Hi-Performance V-8 engine	2,144	210.65
L89	Aluminum cylinder heads	18	394.95
M20	4 speed transmission	28,131	184.80
M21	4 speed close-ratio transmission	11,830	184.80
M22	HD 4 speed transmission	5,410	221.80
M35	Powerglide transmission	106,849	163.70-174.25
M38	300 Deluxe 3 speed automatic transmission	368,871	200.65
M40	3 speed automatic transmission	43,748	221.80
NA9	EEC	65,214	36.90
NK1	Cushioned-rim steering wheel	3,078	34.80
N10	Dual-exhaust system	22,395	30.55
N33	Tilt-type steering wheel	37,760	45.30
N40	Power steering	508,181	105.35
PA3	Special wheel trim cover	56,588	—
PK2	G78-14-B GB white stripe tires	19,604	—
PL3	E78-14-B GB white stripe tires	9,234	26.05
PL4	F70-14-B GB white letter tires	64,758	50.25-65.45
PM6	G78-14-D GB white stripe tires	4,763	—
PU8	G78-14-B GB white stripe tires	114,422	—
PX6	F78-14-B GB white stripe tires	248,753	28.10
PY4	F70-14-B GB white stripe tires	15,341	50.50
PY5	F70-14-B GB red stripe tires	8	—
PY7	G70-14-B GB red stripe tires	2	—
P01	Wheel trim cover	215,176	21.10
P02	Deluxe wheel trim cover	3,358	79.00
P06	Wheel trim ring	3,633	21.10
P90	G70-14-B white stripe tires	26,814	—
P91	G70-14-B GB red stripe tires	194	—

T58	Rear wheel opening skirt	31,442	
T60	HD battery	46,890	15.80
UM1	Push-button AM radio & tape player	27,295	194.85
UM2	Push-button AM/FM stereo radio & tape player	4,532	372.85
U14	Instrument panel gauges	37,378	84.30
U35	Electric clock	59,308	15.80
U46	Lamp monitoring system	5,755	26.35
U63	Push-button radio	487,820	61.10
U69	Push-button AM/FM radio	31,426	133.80
U76	Windshield antenna	580,635	—
U79	AM/FM stereo equipment	7,925	239.10
U80	Auxiliary rear speaker	113,627	13.20
V01	HD radiator	7,347	14.75
V31	Front bumper guard equipment	48,448	15.80
V32	Rear bumper guard equipment	36,595	15.80
V55	Luggage carrier equipment	13,044	—
YD1	Axle for trailering	1,019	10.55
ZJ7	Special wheel, hubcap & trim ring	90,017	35.85
ZJ9	Auxiliary Lighting Group	40,945	21.10– 27.95
ZL2	Special ducted hood air system	28,888	147.45
ZQ9	Performance ratio rear axle	3,348	—
Z15	SS 454 equipment	8,773	503.45
Z20	Monte Carlo SS equipment	3,823	—
Z25	SS 396 equipment	53,599	445.55
795	Vinyl-coated trim	143,031	12.65
961–7	Two-tone color combination	22,130	23.20

*Includes 1970 Monte Carlo.

Facts

Considered the high point in performance, the 1970 Chevelle was also a high point in terms of styling, as it got new front and rear end treatments. The roofline was also modified, and the car did not have side vent windows.

Two Super Sport packages were available in 1970. The Z25 was the SS 396 option and the Z15 was the 454 option. Both packages included the blacked-out grille, domed hood, wheelwell moldings, black rear bumper insert panel, chrome 14x7 five-spoke sport wheels with F70x14 RWL tires, twin rectangular exhaust outlets, SS grille emblem, SS rear bumper emblem, SS emblems on the black steering wheel and column, and—unique to Super Sport models—a black-faced dash panel that came with round gauges. The packages included SS 396 or SS 454 emblems mounted on the front fenders.

Mechanically, the Super Sport came with the F41 suspension, power front disc brakes and either the 402 ci engine (marketed as a 396) or the 454 ci V-8 engine. The standard L34 402 ci was rated at 350 hp and the optional L78 was rated at 375 hp. Only eighteen aluminum-headed L89s were built, as the option was deleted early in the model year.

The big deal was the LS5 and LS6 454 ci engines. Featuring a larger bore and stroke than the 402 (4.25x4.00), the 454 represented the largest production permutation of the Mark IV big-block series. The LS5, rated at 360 hp, came with the small oval-port cylinder heads, Quadrajet carburetor, cast-iron intake manifold and hydraulic camshaft. Available with either the four-speed or Turbo Hydra-matic automatic, an LS5 equipped Chevelle could be optioned with air conditioning. The LS6, rated at 450 hp, came with the large rectangular-port heads, Holley carburetor, aluminum high-rise intake manifold, solid-lifter camshaft and a one-point-higher compression ratio (11.25:1).

Optional on the Super Sport cars was the cowl induction hood. It came with a vacuum-operated rear-facing air valve on the back of the domed hood, which routed outside air into the engine's air cleaner. Cowl induction lettering was used on both sides, and black or white hood and deck Band-Aid stripes were part of the package. The stripes were also available without the cowl induction option. Hood lock pins were included with the Cowl Induction Package.

Conventional antennas were eliminated on all 1970 Chevelles. The radio antenna was now a wire imbedded in the front windshield.

The 1970 Super Sport Chevelles came with clear rather than amber front turn signal lamps.

An additional 402 ci big-block engine was available on all other, non Super Sport, Chevelles. This was the LS3, rated at 330 hp. Installed in 9,338 cars, the engine was known as the Turbo-Jet 400. Cars so equipped could be identified by 400 emblems on the front fenders.

The 1970 Chevelle SS 454 Coupe.

Chapter 17

1971 Chevelle SS and SS 454

Production

2 dr coupe & convertible,
LS5 454 V-8   9,502

2 dr coupe & convertible,
LS3, L48 & L65 V-8s   9,791
Total   19,293

Serial numbers

Description
136371K300001
13637 — Model number (13637-2 dr coupe, 13667-2 dr convertible)
1 — Last digit of model year (1971)
K — Assembly plant (B-Baltimore, K-Leeds, L-Van Nuys, R-Arlington, 1-Oshawa)
300001 — Consecutive sequence number

Location
     On plate attached to driver's side of dash, visible through the windshield.

Engine and transmission suffix codes

CGA — 350 ci V-8 2 bbl 245 hp, manual
CGB — 350 ci V-8 2 bbl 245 hp, Powerglide automatic
CGK, CJJ — 350 ci V-8 4 bbl 270 hp, manual
CGL, CJD — 350 ci V-8 4 bbl 270 hp, Turbo Hydra-matic automatic TH350
CLS — 402 ci V-8 4 bbl 300 hp, HD 3 speed manual
CLA — 402 ci V-8 4 bbl 300 hp, manual
CLB — 402 ci V-8 4 bbl 300 hp, Turbo Hydra-matic automatic
CPA — 454 ci V-8 4 bbl 365 hp, 4 speed manual
CPO — 454 ci V-8 4 bbl 365 hp, Turbo Hydra-matic automatic

Carburetors

350 ci 270 hp — 7041203
350 ci 270 hp w/automatic — 7041202
402 ci 300 hp — 7041201
402 ci 300 hp w/automatic — 7041200
454 ci 365 hp — 7041201
454 ci 365 hp w/automatic — 7041200

Cylinder head casting number

402/454 ci — 3993820

Distributors

350 ci — 1112044
350 ci w/automatic — 1112045
402 ci — 1112057
454 ci — 1112052

Exterior color codes

Antique White	11	Sunflower	52
Nevada Silver	13	Placer Gold	53
Tuxedo Black	19	Sandalwood	61
Ascot Blue	24	Burnt Orange	62
Mulsanne Blue	26	Classic Copper	67
Cottonwood Green	42	Cranberry Red	75
Lime Green	43	Rosewood	78
Antique Green	49		

Interior trim codes

Colors	Cloth bench seats	Vinyl bench seats	Cloth bucket seats	Vinyl bucket seats
Black	704	705	—	706
Dark Blue	725	—	—	—
Dark Jade	730	731	—	732
Sandalwood	718	714	—	715
Dark Saddle	—	721	—	722

Convertible top color codes

White	AA
Black	BB

Vinyl top color codes

White	AA
Black	BB
Blue	CC
Brown	FF
Green	GG

Options

13637 Sport coupe	$2,975.00
13667 Convertible	3,255.00

Option number	Description	Quantity	Retail price
AK1	Deluxe front belts & harness (Chevelle)	91,887	$ 15.30–16.90
AQ2	Electric seatback lock release	195	—
AS4	Deluxe rear seat shoulder harness	245	26.35
AU3	Electric door locks	6,430	46.35–70.60
A01	Tinted glass (all windows)	370,582	43.20
A02	Tinted glass (windshield)	6,827	—
A31	Electric windows	10,774	
A33	Electric tailgate window	18,479	
A39	Custom deluxe front & rear seatbelts	799	13.20–14.75
A41	4 way electric control front seat	1,744	—
A46	4 way electric control front seat	1,186	—
A51	Strato-type front bucket seat	52,638	136.95
A85	Deluxe shoulder harness	72	26.35

A90	Rear compartment lid release	3,060	—
B37	Floor mats	127,473	12.65
B85	Upper bodyside molding	24,174	—
B90	Door & window frame molding	9,547	26.35
B93	Door edge guards	170,733	6.35-9.50
C05	Folding top	5,089	—
C08	Exterior soft trim roof cover	240,635	94.80
C50	Rear window defroster	36,416	31.60-36.90
C51	Rear window air deflector	9,239	—
C60	Deluxe AC	311,464	407.60
D33	Remote control outside mirror	97,133	12.65
D34	Vanity Visor mirror	26,334	3.20
D55	Front compartment floor console	46,614	59.00
D88	Sport stripe	13,382	79.00
F40	Special front & rear suspension	17,711	17.95
F41	Special performance front & rear suspension	16,431	30.55
G67	Rear shock absorber level control	2,297	—
G80	Positraction rear axle	33,994	46.35
JL2	Front disc brakes	83,868	69.55
J50	Vacuum power brake equipment	124,996	47.40
K85	63 amp AC generator	3,620	5.30-26.35
LS3	400 ci V-8 engine	17,656	172.75
LS5	454 ci Hi-Performance V-8 engine	9,502	279.10
L48	350 ci V-8 engine	78,771	73.75
L65	350 ci V-8 engine	119,379	26.35
MC1	HD 3 speed transmission	548	132.00
M11	Floor shift transmission	2,511	132.00
M20	4 speed transmission	9,786	195.40
M22	HD 4 speed transmission	3,035	237.60
M35	Powerglide transmission	66,625	179.55-190.10
M38	3 speed automatic transmission	402,488	216.50
M40	3 speed automatic transmission	21,333	237.60
NK2	Deluxe steering wheel	1,898	15.80
NK4	Sport steering wheel	8,962	15.80
N33	Tilt-type steering wheel	41,853	45.30
N40	Power steering	477,812	115.90

PA3	Special wheel trim cover	24,009	—
PK2	G78-14B GB white stripe tires	16,807	—
PL3	E78-14B 2x2 whitewall tires	153,912	28.15
PM6	G78-14D GB white stripe tires	11,587	—
PM7	F60-15B GB white letter tires	19,542	—
PU8	G78-15B GB white stripe tires	101,248	—
PX6	F78-14B GB white stripe tires	113,593	48.10-53.35
P01	Wheel trim cover	208,162	26.35
P02	Deluxe wheel trim cover	8,392	84.30
P90	G70-15B GB white stripe tires	22,904	—
T58	Rear wheel opening skirt	17,602	—
T60	HD battery	36,319	15.80
UM1	Push-button AM radio & tape player	22,164	200.15
UM2	Push-button AM/FM stereo radio & tape player	3,923	372.85
U14	Instrument panel gauges	16,994	84.30
U35	Electric clock	42,174	16.90
U63	Push-button radio	423,833	66.40
U69	Push-button AM/FM radio	39,079	139.05
U76	Windshield antenna	510,724	—
U79	Stereo equipment	7,894	239.10
U80	Auxiliary speaker	90,512	15.80
V01	HD radiator	5,353	14.75-21.10
V30	Front & rear bumper guards	37,640	31.60
V55	Luggage carrier	17,121	—
YD1	Axle for trailering	1,663	12.65
YF3	Heavy Chevy Package	6,727	
ZJ7	Special wheel, hubcap & trim ring	109,037	45.30
ZJ9	Auxiliary Lighting Group	36,996	15.80-23.70
ZL2	Special ducted hood air	4,079	158.00
ZQ9	Performance ratio rear axle	302	12.65
Z15	SS equipment	19,293	357.05
703	Vinyl interior trim	16,775	—
705	Vinyl interior trim	69,826	—
714	Vinyl interior trim	39,475	—
721	Vinyl interior trim	35,036	—
726	Vinyl interior trim	1,945	—
731	Vinyl interior trim	43,613	—

956	Two-tone color combination	5,991	31.60
960	Two-tone color combination	4,150	31.60
968	Two-tone color combination	3,140	31.60
969	Two-tone color combination	4,511	31.60
971	Two-tone color combination	2,090	31.60
977	Two-tone color combination	3,685	31.60

Facts

The 1971 Chevelles got a new grille with single 7 in. headlights. In the rear, two round taillight lamps were located in each side of the bumper. The center of the front grille and the center of the rear bumper displayed SS emblems. The front turn signal lamps were relocated to the front fender caps. A nice addition was the 15x7 gray mag-style steel wheels. The domed hood was still standard. The interior features were the same, including an SS emblem on the black steering wheel.

Mechanically, the Chevelle was unchanged, save for the engine compartment. Only one 454 ci engine was still available, the LS5, which was uprated to 365 hp even though compression ratio dropped to 9.0:1. Based on Fran Preve's extensive research of the production records for the Tonawanda engine plant where the big-block engines were built, no LS6 engines were released as a production option. The fourteen engines that were built (four tagged CPZ for manual transmission use and ten tagged CPY for automatic use) were never installed in any car.

Besides the LS5 454 ci V-8, three other engines were available on the Chevelle. The L65 245 hp 350 ci small-block V-8 came with a two-barrel carburetor and single exhaust. The L48 350 ci was rated at 270 hp. The LS3 402 ci big-block was rated at 300 hp. The LS3 was marketed as the Turbo-Jet 400.

It cannot be indicated with any certainty how many LS3 engines came with the SS Package, as the big-block, as well as both small-blocks, was available on all other Chevelles with the Super Sport option.

Cowl induction was available in 1971. The optional power steering included a variable-ratio steering box.

The 1971 Chevelle SS 454 Coupe.

Chapter 18

1972 Chevelle SS

Production

2 dr coupe & convertible,
  LS5 454 V-8   5,333

2 dr coupe & convertible,
  LS3, L48 & L65 V-8s   19,613
  Total   24,946

Serial numbers

Description

1D37W2B500001

1 — Chevrolet
D — Chevelle Malibu V-8
37 — Body style (37-2 dr coupe, 67-2 dr convertible)
W — Engine code
2 — Last digit of model year (1972)
B — Assembly plant (B-Baltimore, K-Leeds City, L-Van Nuys, R-Arlington, 1-Oshawa)
500001 — Consecutive sequence number

Engine Codes

F — 307 ci 2 bbl V-8 130 hp
H — 350 ci 2 bbl V-8 165 hp
J — 350 ci 4 bbl V-8 175 hp
U — 402 ci 4 bbl V-8 240 hp
W — 454 ci 4 bbl V-8 270 hp

Engine and transmission suffix codes

CKG — 307 ci V-8 2 bbl 130 hp, manual
CAY — 307 ci V-8 2 bbl 130 hp, manual w/NB2
CKH — 307 ci V-8 2 bbl 130 hp, Powerglide automatic
CAZ — 307 ci V-8 2 bbl 130 hp, Powerglide automatic w/NB2
CTK — 307 ci V-8 2 bbl 130 hp, Turbo Hydra-matic automatic
CMA — 307 ci V-8 2 bbl 130 hp, Turbo Hydra-matic automatic w/NB2
CKA — 350 ci V-8 2 bbl 165 hp, manual
CDA — 350 ci V-8 2 bbl 165 hp, manual w/NB2
CTL — 350 ci V-8 2 bbl 165 hp, Turbo Hydra-matic automatic
CMD — 350 ci V-8 2 bbl 165 hp, Turbo Hydra-matic automatic w/NB2
CKK — 350 ci V-8 4 bbl 175 hp, manual
CDG — 350 ci V-8 4 bbl 175 hp, manual w/NB2
CKD — 350 ci V-8 4 bbl 175 hp, Turbo Hydra-matic automatic
CDD — 350 ci V-8 4 bbl 175 hp, Turbo Hydra-matic automatic w/NB2
CLA, CLS — 402 ci V-8 4 bbl 240 hp, manual
CTA — 402 ci V-8 4 bbl 240 hp, manual
CTH — 402 ci V-8 4 bbl 240 hp, HD 3 speed manual w/AIR
CLB — 402 ci V-8 4 bbl 240 hp, Turbo Hydra-matic automatic
CTB — 402 ci V-8 4 bbl 240 hp, Turbo Hydra-matic automatic w/AIR
CPA — 454 ci V-8 4 bbl 270 hp, manual
CRX — 454 ci V-8 4 bbl 270 hp, manual w/AIR

CRN — 454 ci V-8 4 bbl 270 hp, Turbo Hydra-matic automatic
CRW — 454 ci V-8 4 bbl 270 hp, Turbo Hydra-matic automatic
  w/AIR

Carburetors
402 ci — 7042201
402 ci w/automatic — 7042200
454 ci — 7042201
454 ci W/automatic — 7042200

Distributors
402 ci — 1112057
454 ci — 1112052

Cylinder head casting number
402/454 ci — 6272292

Convertible top color codes
White	AA
Black	BB

Vinyl top color codes
White	AA
Black	BB
Medium Tan	FF
Medium Green	GG
Light Covert	TT

Exterior color codes
Antique White	11
Pewter Silver	14
Ascot Blue	24
Mulsanne Blue	26
Spring Green	36
Gulf Green	43
Sequoia Green	48
Covert Tan	50
Placer Gold	53
Cream Yellow	56
Golden Brown	57
Mohave Gold	63
Orange Flame	65
Midnight Bronze	68
Cranberry Red	75

Interior trim codes
Color	Cloth	Vinyl
Black	703	704
Dark Green	—	713
Medium Tan	—	720
Light Covert	730	732
White	—	743

Options
13637 Sport coupe		$2,980.00
13667 Convertible		3,260.00

Option number	Description	Quantity	Retail price
AK1	Deluxe front belts & harness	112,998	$ 15.30–16.90
AU3	Electric door locks	10,683	46.35–70.60
A01	Tinted glass (all windows)	479,960	43.20
A02	Tinted glass (windshield)	10,965	—
A31	Electric control windows	17,449	—
A33	Electric tailgate window	23,649	—
A39	Custom deluxe front & rear seatbelts	689	13.20–14.75
A41	4 way electric control front seat	3,317	—
A46	4 way electric control front seat	1,831	—

Code	Description	Quantity	Price
A51	Strato-type front bucket seat	61,566	136.95
A85	Deluxe shoulder harness	98	26.35
B37	Floor mats	178,186	12.65
B84	Bodyside molding equipment	143,687	—
B85	Upper bodyside molding	45,201	—
B90	Door & window frame molding	14,570	26.35–33.75
B93	Door edge guards	226,406	6.35–9.50
C05	Folding top	4,852	—
C08	Exterior soft trim roof cover	284,926	94.80
C50	Rear window defroster	45,131	31.60–36.90
C51	Rear window air deflector	10,683	—
C60	Deluxe AC	427,097	407.60
D33	Remote control outside mirror	190,321	12.65
D34	Vanity Visor mirror	44,732	3.20
D55	Front compartment floor console	56,495	59.00
D88	Sport stripe	14,772	79.00
F40	Special front & rear suspension	26,479	17.95
F41	Special performance front & rear suspension	19,595	30.55
G67	Rear shock absorber level control	621	—
G80	Positraction rear axle	40,564	46.35
JL2	Front disc brakes	131,991	69.55
J50	Vacuum power brake equipment	187,151	47.40
K30	Cruise-Master speed control	11,510	63.20
K85	63 amp AC generator	4,918	5.30–26.35
LS3	400 ci V-8 engine	20,031	172.75
LS5	454 ci Hi-Performance V-8 engine	5,333	179.55
L48	350 ci V-8 engine	125,708	73.75
L65	350 ci V-8 engine	274,178	26.35
MC1	HD 3 speed transmission	272	132.00
M11	Floor shift transmission	4,025	132.00
M20	4 speed transmission	10,201	195.40
M22	HD 4 speed transmission	1,513	237.60
M35	Powerglide transmission	43,915	179.55–190.10
M38	3 speed automatic transmission	541,869	216.50
M40	3 speed automatic transmission	21,427	237.60

NK2	Deluxe steering wheel	9,556	15.80
NK4	Sport steering wheel	20,486	15.80
N33	Tilt-type steering wheel	55,301	45.30
N40	Power steering	418,747	115.90
PA3	Special wheel trim cover	24,132	—
PK2	G70-14B G/B white stripe tires	25,929	—
PL3	E78-14B 2x2 whitewall tires	225,900	28.15
PM6	G78-14D G/B white stripe tires	11,012	—
PM7	F60-15B G/B white letter tires	25,695	—
PU8	G78-15B G/B white stripe tires	137,178	—
PX6	F78-14B G/B white stripe tires	78,323	48.10-53.35
P01	Wheel trim cover	223,492	26.35
P02	Deluxe wheel trim cover	18,167	84.30
P90	G70-15B G/B white stripe tires	34,960	—
T58	Rear wheel opening skirt	17,941	—
T60	HD battery	39,907	15.80
UM1	Push-button AM radio & tape player	35,637	200.15
UM2	Push-button AM/FM stereo radio & tape player	7,403	372.85
U14	Instrument panel gauges	25,179	84.30
U35	Electric clock	49,062	16.90
U63	Push-button radio	478,309	66.40
U69	Push-button AM/FM radio	63,234	139.05
U76	Windshield antenna	626,027	—
U79	Stereo equipment	15,366	239.10
U80	Auxiliary speaker	126,641	15.80
V01	HD radiator	36,249	14.75-21.10
V30	Front & rear bumper guards	72,663	31.60
V55	Luggage carrier	26,706	—
YD1	Axle for trailering	2,821	12.65
YF3	Heavy Chevy Package	9,508	142.20
YF5	Calif assembly line emission test	58,689	15.80
ZJ7	Special wheel, hubcap & trim ring	180,852	45.30
ZJ9	Auxiliary Lighting Group	54,605	15.80-23.70
ZL2	Special ducted hood air	3,659	158.00
ZQ9	Performance ratio rear axle	296	12.65
Z15	SS equipment	24,946	357.05
703	Vinyl interior trim	34,259	—

721	Vinyl interior trim	4,814	—
731	Vinyl interior trim	16,958	—
956	Two-tone color combination	6,321	31.60

Facts

Compared with earlier years, 1972 saw the least amount of visual changes on the Chevelle: new front turn signal lamps and a three-tiered grille. All other visual Super Sport features were carried over from 1971: SS emblems on the blacked-out grille, front fenders and rear bumper; domed hood with locking pins; and F60x15 RWL tires on gray 15x7 sport wheels.

Power disc brakes were standard. The F41 suspension was standard only with the 454 ci or 402 ci engines; it could be purchased at additional cost with the other V-8s.

The Super Sport equipment option was now available with any V-8, from the 130 hp 307 ci small-block, to the two 350 ci small-blocks rated at 165 hp and 175 hp, and on to the LS3 402 big-block rated at 240 hp. The LS5 454 ci V-8 was downrated to 270 hp. One reason for lower horsepower ratings was the more realistic SAE net rating system that was adopted industrywide. Another reason was that the engines were retuned to meet harsher emission standards.

A total 1,513 of the 454 ci V-8s were equipped with manual transmission, and 3,820 got the Turbo Hydra-matic automatic. Total 402 and 454 production for the Chevelle was 12,402. The LS3 402, marketed as the Turbo-Jet 400, was also available on non Super Sport equipped Chevelles.

The 1972 Chevelle "Heavy Chevy" and Malibu SS Coupes.

Chapter 19

1973 Chevelle SS

Production
RPO Z15 SS equipment   28,647

Serial numbers
Description
1D37Y3B100001
1 — Chevrolet
D — Chevelle Malibu V-8
37 — Body style (35-4 dr wagon, 37-2 dr coupe)
Y — Engine code
3 — Last digit of model year (1973)
B — Assembly plant (B-Baltimore, K-Leeds, L-Van Nuys,
   R-Arlington, 1-Oshawa)
100001 — Consecutive sequence number

Location
   On plate attached to driver's side of dash, visible through the
windshield.

Engine codes
H — 350 ci 2 bbl V-8 145 hp
J — 350 ci 4 bbl V-8 175 hp
Y — 454 ci 4 bbl V-8 245 hp

Engine and transmission suffix codes
CKA — 350 ci V-8 2 bbl 145 hp, manual
CKC — 350 ci V-8 2 bbl 145 hp, manual w/NB2
CKL — 350 ci V-8 2 bbl 145 hp, Turbo Hydra-matic automatic
CKK — 350 ci V-8 2 bbl 145 hp, Turbo Hydra-matic w/NB2
CKB — 350 ci V-8 4 bbl 175 hp, 4 speed manual
CKH — 350 ci V-8 4 bbl 175 hp, 4 speed manual w/NB2
CKJ — 350 ci V-8 4 bbl 175 hp, Turbo Hydra-matic automatic
CKD — 350 ci V-8 4 bbl 175 hp, Turbo Hydra-matic automatic
   w/NB2
CWA — 454 ci V-8 4 bbl 245 hp, 4 speed manual
CWC — 454 ci V-8 4 bbl 245 hp, 4 speed manual w/NB2
CWB — 454 ci V-8 4 bbl 245 hp, Turbo Hydra-matic automatic
CWD — 454 ci V-8 4 bbl 245 hp, Turbo Hydra-matic automatic
   w/NB2

Carburetors
350 ci 145 hp — 7047311
350 ci 145 hp w/automatic — 7047302
350 ci 175 hp — 7047303
454 ci — 7047316

Distributors

350 ci 145 hp — 1112159
350 ci 175 hp — 1112093

350 ci 175 hp w/automatic —
  1112094
454 ci — 1112113

Exterior color codes

Antique White	11
Tuxedo Black	19
Light Blue	24
Dark Blue	26
Midnight Blue	29
Dark Green	42
Light Green	44
Green/Gold	46
Midnight Green	48
Light Yellow	51
Chamois	56
Light Copper	60
Silver	64
Taupe	66
Dark Brown	68
Dark Red	74
Medium Red	75
Beige	81
Medium Orange	97

Interior trim codes

Black	704
Blue	705
Saddle	720
Chamois	721
Blue/Black	724
Green/Black	730
Light Neutral	732

Vinyl top color codes

Black	—
White	—
Medium Green	—
Medium Blue	—
Light Neutral	—
Chamois	—
Maroon	—

Options

1D35 4 dr 6 passenger Malibu station wagon	$3,290.00
1D35 4 dr 8 passenger Malibu station wagon	3,423.00
1D37 2 dr Malibu Colonade coupe	2,894.00–3,010.00

Option number	Description	Quantity	Retail price
AK1	Custom deluxe belts	184,333	$ 12.50–14.00
AN7	Strato-Bucket seat	127,157	133.00
AU3	Power door lock system	43,715	45.00–69.00
AU6	Power tailgate release	14,587	—
A01	Soft Ray tinted glass	604,433	42.00
A02	Soft Ray tinted windshield	19,387	—
A20	Swing-out rear quarter window	4,336	—
A31	Electric control windows	77,058	75.00–113.00
A42	6 way power seat	15,325	103.00
B37	Color-keyed floor mats	249,064	12.00
B80	Roof drip molding	7,432	—
B84	Bodyside molding	344,787	33.00
B93	Door edge guards	258,317	6.00–9.00
B96	Wheel opening molding	2,717	—

Code	Description	Qty	Price
CA1	Sky roof	9,055	325.00
C08	Vinyl roof cover	286,791	92.00
C50	Rear window defogger	98,607	31.00
C51	Rear window air deflector	12,765	—
C60	Four Season AC	572,288	397.00
D33	Remote control rearview mirror	189,610	12.00
D34	Visor Vanity mirror	34,085	3.00
D35	Sport mirrors	102,178	26.00
D55	Console	116,531	57.00
D88	Sport stripes	6,177	
F40	Special suspension	56,352	17.00
G80	Positraction rear axle	47,055	
J50	Power brakes	348,034	46.00
K30	Speed & cruise control	31,455	
K76	61 amp Delcotron generator	22,904	26.00
K85	63 amp Delcotron generator	882	
LS4	454 ci 4 bbl V–8 engine	22,528	235.00
L48	350 ci 4 bbl V–8 engine	199,984	
L65	350 ci 2 bbl V–8 engine	446,860	
M20	4 speed wide-ratio transmission	3,879	
M21	4 speed close-ratio transmission	1,685	
M38	Turbo Hydra-matic transmission	704,574	—
M40	Turbo Hydra-matic transmission	20,843	
N33	Comfortilt steering wheel	154,750	44.00
N40	Power steering	435,394	113.00
N95	Wire wheel covers	31,757	68.00–82.00
PA3	Deluxe wheel covers	29,713	
PE1	Turbine 1 wheels	22,620	98.00–110.50
PE2	Turbine 2 wheels	3,034	
P01	Full wheel covers	243,631	26.00
QCF	G70x14 B white letter tires	26,934	61.75
QEH	E78x14 B white stripe tires	7,769	28.00
QGL	G78x14 B white stripe tires	258,608	32.00
QGT	G78x15 B white stripe tires	3,805	
QHF	H78x14 B white stripe tires	56,771	58.65
QRM	GR70x15 steel belt radial white stripe tires	271,254	—
T60	HD battery	70,120	15.00
UM1	Stereo tape player w/AM radio	55,520	195.00
UM2	Stereo tape player w/stereo radio	22,193	363.00
U14	Special instrumentation	51,564	82.00

U35	Electric clock	61,037	16.00
U58	AM/FM stereo radio	48,035	233.00
U63	Push-button AM radio	456,294	65.00
U69	Push-button AM/FM radio	111,539	135.00
U76	Windshield antenna	540	—
U80	Rear seat speaker	173,598	15.00
VE5	Deluxe bumpers	71,397	24.00
V01	HD radiator	89,399	14.00–21.00
V30	Bumper guards	190,693	31.00
V55	Roof carrier	34,309	—
YA2	El Camino Estate equipment	6,723	—
YA7	Calif assembly line emission test	71,346	15.00
YD1	Trailering axle ratio	6,276	—
YJ9	Exterior Decor Package	125,067	14.04–26.52
ZJ7	Rally wheels	85,859	
ZJ9	Auxiliary lighting	103,859	12.48–18.33
ZQ9	Performance axle ratio	6,898	—
Z15	SS equipment	28,647	242.75

Facts

All General Motors intermediates were totally restyled for 1973. The Chevelle, based on a 113 in. wheelbase, looked bigger than ever. The Super Sport option was available on the two-door Malibu coupe and on the two- and three-seat wagons.

The Super Sport option consisted of a blacked-out grille with SS emblem, dual sport mirrors, SS fender and rear panel emblems, lower bodyside and wheel opening striping, black accented taillight bezels, and bright rear quarter window and drip moldings. In the interior, Super Sport equipped Malibus got a special instrument cluster with black bezels and SS emblems on the steering wheel and door panels.

Also standard were a rear stabilizer bar, 14x7 rally wheels and G70x14 RWL tires.

Two small-blocks—the L65 145 hp and L48 175 hp 350 ci V-8s—and the LS4 454 big-block rated at 245 hp were all that was available.

The year 1973 was the last for the Super Sport option on the Chevelle.

The 1973 Chevelle SS Coupe.

The 1967 Camaro SS 350 Coupe.

Chapter 20

1967 Camaro SS 350 and SS 396

Production

8 cyl	
12437 2 dr coupe	142,242
12467 2 dr convertible	19,856
Total	162,098

SS equipment option
RPO L48 SS w/295 hp
350 ci   29,270

RPO L35 SS w/325 hp
396 ci   4,003
RPO L78 SS w/375 hp
396 ci   1,138
Total   34,411

Serial numbers

Description
124377L100001
12437—Model number (12437-2 dr coupe, 12467-2 dr convertible)
7 — Last digit of model year (1967)
L — Assembly plant (L-Los Angeles, N-Norwood)
100001 — Consecutive sequence number

Location
    On plate attached to left front door hinge post.

Engine and transmission suffix codes
MS — 350 ci V-8 4 bbl 295 hp, 3 or 4 speed manual
MT — 350 ci V-8 4 bbl 295 hp, 3 or 4 speed manual w/AIR
MU — 350 ci V-8 4 bbl 295 hp, Powerglide automatic
MV — 350 ci V-8 4 bbl 295 hp, Powerglide automatic w/AIR
MW — 396 ci V-8 4 bbl 325 hp, manual or Turbo Hydra-matic
    automatic
MX — 396 ci V-8 4 bbl 325 hp, 3 or 4 speed manual w/AIR
MY — 396 ci V-8 4 bbl 325 hp, Turbo Hydra-matic automatic
MZ — 396 ci V-8 4 bbl 325 hp, Turbo Hydra-matic automatic w/AIR
EI — 396 ci V-8 4 bbl 350 hp, 3 or 4 speed manual
EY — 396 ci V-8 4 bbl 350 hp, 3 or 4 speed manual w/AIR
EQ — 396 ci V-8 4 bbl 350 hp, Turbo Hydra-matic automatic
EZ — 396 ci V-8 4 bbl 350 hp, Turbo Hydra-matic automatic w/AIR
MQ — 396 ci V-8 4 bbl 375 hp, 4 speed manual
MR — 396 ci V-8 4 bbl 375 hp, 4 speed manual w/AIR

Carburetors
396 ci 325 hp — 7027201
396 ci 325 hp — 7027200

396 ci 350 hp — 3908957 (Holley R–3837A)
396 ci 350 hp w/AIR — 3908959 (Holley R–3839A)
396 ci 350 hp w/Turbo Hydra-matic — 3908956 (Holley R–3836A)
396 ci 350 hp w/AIR & Turbo Hydra-matic — 3908958
   (Holley R–3838A)
396 ci 375 hp — 3916143 (Holley R–3910A)
396 ci 375 hp w/AIR — 3916145 (Holley R–3911A)

Distributors

350 ci — 1111168	396 ci 350 hp — 1111170
396 ci — 1111169	396 ci 375 hp — 1111170

Exterior color codes

Tuxedo Black	AA	Emerald Turquoise	KK
Ermine White	CC	Tahoe Turquoise	LL
Nantucket Blue	DD	Royal Plum	MM
Deepwater Blue	EE	Madeira Maroon	NN
Marina Blue	FF	Bolero Red	RR
Granada Gold	GG	Sierra Fawn	SS
Mountain Green	HH	Capri Cream	TT
		Butternut Yellow	YY

Interior trim codes

Color	Std bucket seats	Bench seats	Custom bucket seats
Black	760	756	765
Red	741	—	742
Gold	709	796	711
Blue	717	739	—
Turquoise	—	—	779
Bright Blue	—	—	732
Parchment/Black	—	—	797
Yellow	—	—	707

Convertible top color codes

White	AA
Black	BB
Medium Blue	—

Vinyl top color codes

Black	BB
Light Fawn	—

Options

12437 Sport coupe, 327 ci 210 hp V-8	$2,572.00
12467 Convertible, 327 ci 210 hp V-8	2,809.00

Option number	Description	Quantity	Retail price
AL4	Strato-Back front seat (NA w/convertible or center console)	6,583	$ 26.35
AS1	Front shoulder belts	477	23.20
AS2	Strato-Ease headrests	2,342	52.70
A01	Tinted glass, all windows	34,725	30.55
A02	Tinted glass, windshield	81,998	21.10

A31	Power windows	4,957	100.10
A39	Custom deluxe front & rear seatbelts	51,247	6.35
A67	Folding rear seat	17,993	31.60
A85	Custom deluxe belts	894	26.35
B37	Color-keyed floor mats	23,747	10.55
B93	Door edge guards	37,964	3.20
C06	Power convertible top (white, black or blue)	11,783	52.70
C08	Vinyl roof cover (black or beige; sport coupe only)	52,455	73.75
C48	Heater & defroster deletion	2,201	31.65 CR
C50	Rear window defroster (sport coupe only)	7,031	42.15
C60	AC	28,226	356.00
C80	Positraction rear axle	31,792	42.15
C94	Non-std 3.31:1 ratio axle	1,177	2.15
C96	Non-std 3.55:1 ratio axle	6,322	2.15
C97	Non-std 2.73:1 ratio axle	539	2.15
D33	Remote control outside LH mirror	8,630	9.50
D55	Console	129,477	47.40
D91	Front end accent band	24,370	14.75
F41	Special purpose front & rear suspension	5,968	10.55
H01	Non-std 3.07:1 ratio axle	560	2.15
H05	Non-std 3.73:1 ratio axle	2,281	2.15
J50	Power brakes	24,549	42.15
J52	Front disc brakes	14,899	79.00
J56	HD front disc brakes w/metallic brakes (Z28 required)	205	105.35
J65	Special brakes w/metallic facings	1,217	36.90
K02	Temperature-controlled fan (V–8 only; incl w/AC & 302 ci engine)	2,375	15.80
K19	GM AIR	34,096	44.75
K24	Closed positive engine ventilation	34,503	5.25
K30	Cruise-Master speed & cruise control (automatic transmission required; V–8 only)	305	50.05
K76	61 amp Delcotron generator (incl w/AC)	136	21.10
K79	42 amp Delcotron generator (NA w/AC)	362	10.55
L22	155 hp Turbo-Thrift 250 ci 6 cyl engine	38,165	26.35
L30	275 hp Turbo-Fire 327 ci V–8 engine	25,287	92.70

L35	Camaro SS w/325 hp Turbo-Jet 396 ci engine	4,003	263.30
L48	Camaro SS w/295 hp Turbo-Fire 350 ci engine	29,270	210.65
L78	Camaro SS w/375 hp Turbo-Jet 396 ci engine	1,138	500.30
M11	Floor-mounted shift lever	12,051	10.55
M13	3 speed transmission (350 ci engine required)	681	79.00
M20	4 speed wide-range transmission (NA w/302 ci or 375 hp 396 ci engine)	45,806	184.35
M21	4 speed close-ratio transmission (375 hp 396 ci engine required)	1,733	184.35
M35	Powerglide automatic transmission (NA w/396 ci engine)	122,727	194.85
M40	Turbo Hydra-matic transmission (325 hp engine required)	1,453	226.45
N10	Dual-exhaust system	6,722	21.10
N30	Deluxe steering wheel ($4.25 w/custom interior)	30,967	7.40
N33	Comfortilt steering wheel (opt transmission or console required)	7,980	42.15
N34	Sports-styled walnut-grained plastic steering wheel	8,065	31.60
N40	Power steering	92,181	84.30
N44	Special steering (requires power steering when AC or 325 hp engine is ordered)	6,155	15.80
N61	Dual-exhaust system (for 210 hp or 275 hp engine; NC for SS models)	13,748	21.10
N96	4 mag-style wheel covers (NA w/disc brakes)	6,630	73.75
PQ2	7.35x14 2 ply nylon white stripe tires (NC w/SS models)	10,913	52.00
PW6	D70x14 2 ply nylon red stripe tires (std w/SS models)	8,330	62.50
P01	4 brightmetal wheel covers (NA w/disc brakes)	137,163	21.10

Code	Description	Qty	Price
P02	4 simulated wire wheel covers (NA w/disc brakes)	6,577	73.75
P12	5 14x6 wheels	3,667	5.30
P58	7.35x14 2 ply whitewall tires	138,998	31.35
T60	HD battery	7,964	7.40
U03	Tri-Volume horn (sport coupe only)	1,580	13.70
U15	Speed warning indicator	3,698	10.55
U17	Special instrumentation	27,078	79.00
U25	Luggage compartment light	24,787	2.65
U26	Underhood light	22,965	2.65
U27	Glove compartment light	11,032	2.65
U28	Ashtray light	25,538	1.65
U29	Courtesy lights (sport coupe only)	23,691	4.25
U35	Floor-mounted electric clock	13,185	15.80
U57	Stereo tape system	2,746	128.50
U63	Push-button AM radio	174,021	57.40
U69	Push-button AM/FM radio	6,232	133.80
U73	Manual rear antenna (NA w/AM/FM radio)	32,223	9.50
U80	Rear seat speaker (NA w/stereo tape system)	27,701	13.20
V01	HD radiator (incl w/AC; NA w/325 hp or 375 hp engine)	6,190	10.55
V31	Front bumper guard	35,154	12.65
V32	Rear bumper guard	35,029	9.50
Z21	Style Trim Group	79,016	40.05
Z22	Rally Sport Package	64,842	105.35
Z23	Special Interior Group	74,648	10.55
Z28	Special Performance Package (HD front disc brakes w/metallic rear brakes; Positraction recommended; sport coupe V-8 only)	602	328.10
	W/plenum air intake incl special air cleaner & duct systems	—	437.10
	W/exhaust headers	—	779.40
	W/plenum air intake & exhaust headers	—	858.40
Z87	Custom interior	69,103	94.80

Facts

By far, the most popular performance Camaro models in the car's first year of production were the Super Sport models.

The SS Option Package was available on both the coupe and convertible Camaros. The cars used SS 350 emblems in the center of the grille and just SS emblems on the front of both front fenders. On the hood, two simulated oil cooler louvers set the Super Sport apart from regular Camaros. An SS 350 emblem also appeared on the fuel filler cap, which was tethered. The distinctive bumblebee stripes, in black or white, surrounded the Camaro's nose. Other Super Sport features were standard Red Line tires and underhood insulation.

All 396 ci powered Camaros came with SS emblems mounted on the grille and gas cap and the distinctive 396 V insignias on the front fenders behind the wheelwells beneath the Camaro emblem. The 396 ci Camaros also came with a black-painted rear taillight panel.

In the interior, an SS emblem was displayed on the steering wheel.

Three engines were available. The 295 hp version of the small-block V-8 displaced 350 ci. First used on the 1967 Camaro, the 350 came with a Quadrajet four-barrel carburetor; hydraulic cam; dual exhausts; and chrome air cleaner, valve covers and oil filler cap.

Optional were two versions of the Mark IV big-block 396 ci engines. The first, rated at 325 hp, came with the small oval-port heads, cast-iron intake manifold, Quadrajet carburetor and hydraulic cam. The second, the L78 375 hp 396 ci, came with big-port cylinder heads, aluminum high-rise intake manifold, Holley carburetor and solid-lifter camshaft.

The Rally Sport Package, RPO Z22, was available in conjunction with the SS Package, giving the Camaro a different, distinctive look. Most noticeable were the hidden, electrically operated headlights. Because the grille-mounted turn signal lamps were eliminated, the turn signal lights were relocated to the lower valance panel. Also included were bodyside stripes, rocker panel moldings, wheelwell moldings, roof drip rail moldings for the coupe and black-painted rear taillight bezels. Taillamps were all-red lenses. When the Rally Sport Package was installed with the SS Package, the RS emblems normally used on the grille fender and gas cap were superseded by SS emblems.

Disc brakes were optional, but the slotted rally wheels were required.

All 1967 Super Sport Camaros came with monoleaf rear springs and a traction bar on the right rear.

The bumblebee nose stripes were made available on other, non Super Sport equipped Camaros in March 1968.

Standard with the Super Sport models was the F41 Suspension Package, which included heavy-duty springs and shocks.

The new Camaro was chosen to be the pace car at the 1967 Indy 500. Four were built for use in the race and about 100 additional examples for use by race officials and other dignitaries. All were white Super Sport, Rally Sport convertibles with Bright Blue custom interiors. No breakdown exists on how many got the 350 ci and 396 ci engines.

Chapter 21

1968 Camaro SS

Production

8 cyl

12437 2 dr coupe	167,251
12467 2 dr convertible	16,927
Total	184,178

SS equipment option

RPO L48 SS w/295 hp 350 ci	12,496
RPO L35 SS w/325 hp 396 ci	10,773
RPO L34 SS w/350 hp 396 ci	2,579
RPO L78 SS w/375 hp 396 ci	4,575
RPO L89 SS w/375 hp 396 ci	272
Total	30,695

Serial numbers

Description

124378L100001
12437 — Model number (12437-2 dr coupe, 12467-2 dr convertible)
8 — Last digit of model year (1968)
L — Assembly plant (L-Los Angeles, N-Norwood)
100001 — Consecutive sequence number

Location

On plate attached to driver's side of dash, visible through the windshield.

Engine and transmission suffix codes

MS — 350 ci V-8 4 bbl 295 hp, 3 or 4 speed manual
MU — 350 ci V-8 4 bbl 295 hp, Powerglide automatic
MW — 396 ci V-8 4 bbl 325 hp, manual or Turbo Hydra-matic automatic
MY — 396 ci V-8 4 bbl 325 hp, Turbo Hydra-matic automatic
MX — 396 ci V-8 4 bbl 350 hp, 4 speed manual
MR — 396 ci V-8 4 bbl 350 hp, Turbo Hydra-matic automatic
MQ — 396 ci V-8 4 bbl 375 hp, 4 speed manual
MT — 396 ci V-8 4 bbl 375 hp, 4 speed manual w/aluminum heads

Carburetors

396 ci — 7028211
396 ci W/automatic — 7028210
396 ci 350 hp — 7028217
396 ci 350 hp W/automatic — 7028218
396 ci 375 hp — 3923289 (Holley R-4053A)

Distributors

396 ci — 1111169
396 ci 350 hp — 1111145
396 ci 350 hp w/automatic — 1111169
396 ci 375 hp — 1111170

Exterior color codes

Tuxedo Black	AA	Cordovan Maroon	NN
Ermine White	CC	Corvette Bronze	OO
Grotto Blue	DD	Seafrost Green	PP
Fathom Blue	EE	Matador Red	RR
Island Teal	FF	Palomino Ivory	TT
Ash Gold	GG	LeMans Blue	UU
Grecian Green	HH	Sequoia Green	VV
Rallye Green	JJ	Butternut Yellow	YY
Tripoli Turquoise	KK	British Green	ZZ
Teal Blue	LL		

Interior trim codes

Color	Std bench seats	Deluxe bench seats	Std bucket seats	Deluxe bucket seats
Black	714	715	712	713
Blue	719	720	717	718
Gold	723	—	721	722
Red	—	—	724	725
Turquoise	—	727	—	726
Parchment/Black	—	—	—	711
Ivory Houndstooth	—	—	—	716
Black Houndstooth	—	—	—	749

Convertible top color codes

White	AA
Black	BB
Blue	—

Vinyl top color codes

Black	BB
White	AA

Options

12437 Sport coupe	$2,670.00
12467 Convertible	2,908.00

Option number	Description	Quantity	Retail price
AK1	Custom deluxe seatbelts and shoulder belts	22,988	$ 11.10
AL4	Strato-Back front seat (NA w/convertible or center console)	4,896	32.65
AS1	Front shoulder belts	70	23.20
AS2	Head restraints (driver & passenger)	2,234	52.70
AS4	2 custom-deluxe-type rear shoulder belts	109	26.35
AS5	2 std-type rear shoulder belts	24	23.20
A01	Soft Ray tinted glass (all windows)	65,329	30.55

A02	Soft Ray tinted glass (windshield only)	60,677	21.10
A31	Power windows	3,304	100.10
A39	Custom deluxe front & rear seatbelts (convertible only)	3,560	7.90
A67	Folding rear seat	7,384	42.15
A85	Custom deluxe front shoulder belts (convertible only)	222	26.35
B37	2 front & 2 rear color-keyed floor mats	30,713	10.55
B93	Door edge guards	49,395	4.25
C06	Power convertible top	9,580	52.70
C08	Vinyl roof cover	77,065	73.75
C50	Rear window defroster ($31.60 for convertible)	6,181	21.10
C60	Four Season AC (NA w/375 hp engine)	35,866	360.20
C80	Positraction rear axle	36,701	42.15
D33	Remote control outside LH mirror	4,740	9.50
D55	Console	140,530	50.60
D80	Auxiliary panel & valance (spoiler)	15,520	32.65
D90	Sport striping (incl w/SS option)	30,541	25.30
D91	Front end accent band (NA w/Camaro SS or Z28)	40,487	14.75
F41	Special purpose front & rear suspension	7,117	10.55
G31	Special rear springs	821	20.05
J50	Vacuum power drum brakes	44,196	42.15
J52	Vacuum power disc front brakes	20,117	100.10
KD5	HD closed positive engine ventilation	30	6.35
K02	Temperature-controlled fan	1,285	15.80
K30	Cruise-Master speed control (automatic transmission required; V-8 only)	327	52.70
K76	HD generator ($5.30 w/AC)	90	26.35
K79	42 amp Delcotron generator	189	10.55
L30	275 hp Turbo-Fire 327 ci V-8 engine	21,686	92.70
L34	Camaro SS w/350 hp Turbo-Jet 396 ci engine	2,579	368.65

L35	Camaro SS w/325 hp Turbo-Jet 396 ci V-8 engine	10,773	263.30
L48	Camaro SS w/295 hp Turbo-Fire 350 ci V-8 engine	12,496	210.65
L78	Camaro SS w/375 hp Turbo-Jet 396 ci V-8 engine	4,575	500.30
L89	Camaro SS w/375 hp Turbo-Jet 396 ci V-8 engine & aluminum cylinder heads	272	868.95
M11	Floor-mounted shift lever	30,192	10.55
M13	Special 3 speed transmission	752	79.00
M20	4 speed wide-range transmission (NA w/L89)	35,161	184.35
M21	4 speed close-ratio transmission (requires 350 hp or 375 hp engine)	11,134	184.35
M22	4 speed close-ratio transmission (requires 375 hp engine)	1,277	310.70
M35	Powerglide automatic transmission (NA w/396 ci engines)	127,165	194.85
M40	Turbo Hydra-matic transmission	5,466	237.00
NF2	Dual-exhaust system, a low tone muffler	9,024	27.40
N10	Dual exhaust equipment	4,462	27.40
N30	Deluxe steering wheel	9,178	4.25
N33	Comfortilt steering wheel	5,294	42.15
N34	Sports-styled wood-grain steering wheel w/plastic rim	5,649	31.60
N40	Power steering	115,280	84.30
N44	Special steering	3,090	15.80
N65	Space Saver spare tire	1,021	19.35
N95	Simulated wire wheel covers (NA w/Z28)	3,988	73.75
N96	Mag-style wheel covers (NA w/Z28)	6,072	73.75
PA2	Mag-spoke wheel covers (NA w/Z28)	4,085	73.75
PW7	F70x14 2 ply white stripe tires (NC w/SS models)	26,670	64.75
PW8	F70x14 2 ply red stripe tires (NC w/SS models)	6,686	64.75

PY4	F70x14 2 ply belted white stripe tires ($26.55 w/SS models)	—	26.55
PY5	F70x14 2 ply belted red stripe tires ($26.55 w/SS models)	—	26.55
P01	Brightmetal wheel covers (NA w/Z28)	133,742	21.10
P58	7.35x14 2 ply original equipment whitewall tires	141,178	31.35
T60	HD battery	8,196	7.40
U03	Tri-Volume horn	768	13.70
U15	Speed warning indicator	2.344	10.55
U17	Special instrumentation	20,263	94.80
U35	Electric clock (incl w/special instrumentation)	20,319	15.80
U46	Light monitoring system	1,755	26.35
U57	Stereo tape system	4,155	133.80
U63	Push-button AM radio	192,805	61.10
U69	Push-button AM/FM radio	7,214	133.80
U73	Manual rear antenna	21,729	9.50
U79	Push-button AM/FM stereo radio	1,335	239.15
U80	Rear seat speaker (NA w/stereo)	23,198	13.20
V01	HD radiator (incl w/AC; NA w/302 ci or 396 ci engine)	4,682	13.70
V31	Front bumper guard	19,455	12.65
V32	Rear bumper guard	18,628	12.65
ZJ7	Rally wheels (incl special wheel, hubcap & trim ring)	8,047	31.60
ZJ9	Auxiliary Lighting Group A	18,099	13.70
Z21	Style Trim Group	93,235	42.15
Z22	Rally Sport Package	40,977	105.35
Z23	Special interior	57,098	17.95
Z28	Special Performance Package	7,199	400.25
Z87	Custom interior	50,461	110.60
GRP1	Appearance Guard Group All models wo/special rear springs	—	40.10
	All models w/special rear springs	—	27.45
GRP4	Operating Convenience Group Sport coupe w/special instrumentation	—	30.60

Sport coupe wo/special instrumentation	—	46.40
Convertible w/special instrumentation	—	9.50
Convertible wo/special instrumentation	—	23.30
— Optional axle ratios	—	2.15

Facts

The 1968 Camaro got several styling changes. Side vent windows were no longer available, side marker lights were included on all Camaros thanks to government regulations and the grille had a pointed center. Different taillight bezels with two lamps per side were used. Cars without the RS Package got lenses in red and white; cars with the Rally Sport got both lenses in red and the back-up lights relocated on the lower valance panel.

The SS Package, available on the coupe and convertible, was slightly modified. The center of the front grille and the gas cap carried SS emblems without engine numerals. The front fenders also used SS emblems, behind the wheelwells just below the Camaro emblem. Hoods with the 350 ci engine were unchanged; those with the 396 ci engine got eight simulated carburetor stacks. Super Sport Camaros with the 396 engine also had the rear taillight panel painted black. The same bumblebee front stripe from earlier years was used; however, later in the year, a variation of the stripe appeared, which continued along the side of the car, ending at the rear of the door.

Super Sport models could also be equipped with the RS Package. As with 1967, the Rally Sport models used a different grille that hid the front headlights and required relocated turn signal lamps—on the lower valance panel. The headlight doors were vacuum operated. Other features of the RS Package included wheelwell moldings, drip moldings and lower body moldings with black-painted rocker panels.

No SS emblems were used in the interior.

Several mechanical improvements were made. All Super Sport Camaros got a multileaf rear spring suspension replacing the monoleaf setup of 1967 and staggered rear shocks, which meant the passenger's side rear shock was located behind the axle housing and the driver's-side rear shock was located in front. The use of staggered shocks eliminated wheel hop. Super Sport Camaros also got finned front brake drums.

Standard Super Sport engine was the same 295 hp 350 ci V-8. The L35 and L78 396 big-blocks were carried over and joined by the L35 and L89 versions of the 396 as well. The L35 396 was rated at 350 hp and the L89 396 was the same as the L78, save for the aluminum cylinder heads.

Rear spoilers, often seen on Z-28 Camaros, were available on other Camaros as well under option D80.

Chapter 22

1969 Camaro SS

Production

8 cyl		**SS equipment option**	
12437 2 dr coupe	190,971	RPO Z27	34,932
12467 2 dr convertible	15,866		
Total	206,837		

Serial numbers

Description
124379L500001
12437 — Model number (12437–2 dr coupe, 12467–2 dr convertible)
9 — Last digit of model year (1969)
L — Assembly plant (L–Los Angeles, N–Norwood)
500001 — Consecutive sequence number

Location
On plate attached to driver's side of dash, visible through the windshield.

Engine and transmission suffix codes

HA — 350 ci V-8 4 bbl 300 hp, 3 or 4 speed manual
HP — 350 ci V-8 4 bbl 300 hp, manual w/10 in. clutch
HE — 350 ci V-8 4 bbl 300 hp, Powerglide automatic
HB — 350 ci V-8 4 bbl 300 hp, Turbo Hydra-matic automatic
JU — 396 ci V-8 4 bbl 325 hp, 4 speed manual
CJU — 402 ci V-8 4 bbl 325 hp, 4 speed manual
JB — 396 ci V-8 4 bbl 325 hp, 4 speed manual w/HD clutch
CJB — 402 ci V-8 4 bbl 325 hp, 4 speed manual w/HD clutch
JG — 396 ci V-8 4 bbl 325 hp, Turbo Hydra-matic automatic
CJG — 402 ci V-8 4 bbl 325 hp, Turbo Hydra-matic automatic
JF — 396 ci V-8 4 bbl 350 hp, 4 speed manual
CJF — 402 ci V-8 4 bbl 350 hp, 4 speed manual
KA — 396 ci V-8 4 bbl 350 hp, 4 speed manual w/HD clutch
CKA — 402 ci V-8 4 bbl 350 hp, 4 speed manual w/HD clutch
JI — 396 ci V-8 4 bbl 350 hp, Turbo Hydra-matic automatic
CJI — 402 ci V-8 4 bbl 350 hp, Turbo Hydra-matic automatic
JH — 396 ci V-8 4 bbl 375 hp, 4 speed manual
CJH — 402 ci V-8 4 bbl 375 hp, 4 speed manual
KC — 396 ci V-8 4 bbl 375 hp, 4 speed manual w/HD clutch
CKC — 402 ci V-8 4 bbl 375 hp, 4 speed manual w/HD clutch
JJ — 396 ci V-8 4 bbl 375 hp, 4 speed manual w/aluminum heads
CJJ — 402 ci V-8 4 bbl 375 hp, 4 speed manual w/aluminum heads
KE — 396 ci V-8 4 bbl 375 hp, 4 speed manual w/HD clutch & aluminum heads

CKE — 402 ci V-8 4 bbl 375 hp, 4 speed manual w/HD clutch & aluminum heads
JL — 396 ci V-8 4 bbl 375 hp, Turbo Hydra-matic automatic
CJL — 402 ci V-8 4 bbl 375 hp, Turbo Hydra-matic automatic
JM — 396 ci V-8 4 bbl 375 hp, Turbo Hydra-matic automatic w/aluminum heads
CJM — 402 ci V-8 4 bbl 375 hp, Turbo Hydra-matic automatic w/aluminum heads

Carburetors
396 ci — 7029215
396 ci w/automatic — 7029204
396 ci 350 hp — 7029215
396 ci 350 hp w/automatic — 7029204
396 ci 375 hp — 3959164 (Holley R-4346A)

Distributors
396 ci — 1111489
398 ci w/AT — 1111497
396 ci 350 hp — 1111499
396 ci 375 hp — 1111499

Two-tone color codes*

Glacier Blue/Dover White	53/50
Glacier Blue/Dusk Blue	53/51
Dusk Blue/Glacier Blue	51/53
Olympic Gold/Dover White	65/50
Burnished Brown/Champagne	61/63
Azure Turquoise/Dover White	55/50
*Lower/upper.	

Exterior color codes

Tuxedo Black	10
Butternut Yellow	40
Dover White	50
Dusk Blue	51
Garnet Red	52
Glacier Blue	53
Azure Turquoise	55
Fathom Green	57
Frost Green	59
Burnished Brown	61
Champagne	63
Olympic Gold	65
Burgundy	67
Cortez Silver	69
LeMans Blue	71
Hugger Orange	72
Daytona Yellow	76
Rallye Green	79

Interior trim codes

Color	Std	Custom
Black	711	712
Blue	715	716
Dark Green	723	722
Midnight Green	721	725
Red	718	719
Black Houndstooth	—	713
Ivory Houndstooth	—	729
Yellow Houndstooth	—	714
Orange Houndstooth	—	720

Convertible top color codes

White	AA
Black	BB

Vinyl top color codes

Black	BB
Parchment	EE
Dark Blue	CC
Dark Brown	FF
Midnight Green	GG

Options

12437 Sport coupe	$2,727.00
12467 Convertible	2,940.00

Option number	Description	Quantity	Retail price
AS1	Front shoulder belts (convertible only)	47	$ 23.20
AS4	Custom-deluxe-type rear shoulder belts	37	26.35
AS5	Rear shoulder belts (sport coupe)	78	23.20
A01	Tinted glass (all windows)	114,733	32.65
A31	Power windows	3,058	105.35
A39	Custom deluxe front & rear seatbelts (convertible only)	4,901	9.00
A67	Folding rear seat	4,397	42.15
A85	Custom deluxe front shoulder belts (convertible only)	922	26.35
B37	Front & rear color-keyed mats	37,158	11.60
B93	Door edge guards	57,128	4.25
CE1	Headlight washer (incl w/Z22)	116	15.80
C06	Power convertible top	9,631	52.70
C50	Forced air rear window defroster ($32.65 for convertible)	7,912	22.15
C60	AC (NA w/375 hp engine)	44,737	376.00
DX1	Front accent stripping (NA w/Camaro SS or D90)	20,479	25.30
D33	Remote control outside LH mirror	7,771	10.55
D34	Vanity Visor mirror	9,00	3.20
D55	Console	156,225	53.75
D80	Spoiler (incl w/Z28)	19,040	32.65
D90	Sport striping (incl w/Z27)		
D96	Fender striping	5,176	15.80
F41	Special purpose front & rear suspension	5,929	10.55
G31	Special rear springs (incl rear bumper guards)	556	20.05
G80	Positraction rear axle	48,755	42.15

JL8	4 wheel power disc brakes (incl F70x15 belted white stripe tires w/Z27; $623.59 wo/Z27 or Z28)	206	500.30
J50	power drum brakes	82,890	42.15
J52	Power disc brakes (incl w/Z27)	67,231	64.25
KD5	HD closed positive engine ventilation	52	6.35
K02	Temperature-controlled fan (incl 375 hp engine)	1,052	15.80
K05	Engine block heater	2,124	10.55
K79	42 amp Delcotron generator (NA w/375 hp engine, Z28 or AC)	224	10.55
K85	63 amp Delcotron generator (NA w/375 hp engine or Z28; $5.30 w/AC)	114	26.35
LM1	255 hp Turbo-Fire 350 ci V-8 engine	10,406	52.70
L34	350 hp Turbo-Jet 396 ci V-8 engine (requires Z27)	2,018	184.35
L35	325 hp Turbo-Jet 396 ci V-8 engine (requires Z27)	6,752	63.20
L48	300 hp Turbo-Fire 350 ci V-8 engine (incl w/Z27)	22,339	—
L65	250 hp Turbo-Fire 350 ci V-8 engine	26,898	21.10
L78	375 hp Turbo-Jet 396 ci V-8 engine (requires Z27)	4,889	316.00
L89	375 hp Turbo-Jet 396 ci V-8 engine w/aluminum cylinder heads (requires Z27)	311	710.95
MC1	3 speed transmission (requires 300 hp or 396 ci engine)	3,079	79.00
M20	4 speed wide-ratio transmission	37,816	195.40
M21	4 speed close-ratio transmission (requires 302 ci, 350 hp or 375 hp engine);	26,501	195.40
M22	HD 4 speed close-ratio transmission (requires 375 hp engine)	2,117	322.10
M35	Powerglide automatic transmission (w/V-8s exc 396 ci or 302 ci engines)	78,849	174.25

Code	Description	Qty	Price
M40	Turbo Hydra-matic automatic transmission (w/325 hp or 350 hp engines; $290.40 w/375 hp Z27 engine)	66,423	221.80
NC8	Dual-chambered-exhaust system	1,526	15.80
N10	Dual-exhaust system (incl w/Z27 or Z28)	5,545	30.55
N33	Comfortilt steering wheel	6,575	45.30
N34	Sports-styled wood-grain steering wheel w/plastic rim	6,883	34.80
N40	Power steering (incl quick-ratio steering w/Z27)	141,607	94.80
N44	Special steering	2,161	15.80
N65	Space Saver spare tire	2,228	19.00
N95	Simulated wire wheel covers	2,118	73.75
N96	Mag-style wheel covers	2,866	73.75
PA2	Mag-spoke wheel covers	1,362	73.75
PK8	E78x14 2 ply whitewall tires	102,328	32.10
PL5	F70x14 2 ply white letter tires	30,605	63.05
PW7	F70x14 2 ply white stripe tires	14,457	NC
PW8	F70x14 2 ply red stripe tires	6,243	NC
PY4	F70x14 fiberglass belt white stripe tires	5,783	26.25
PY5	F70x14 fiberglass belt red stripe tires	1,085	26.25
P01	Brightmetal wheel covers	106,386	21.10
P06	Wheel trim rings (for use w/ std hubcaps only)	2,401	21.10
T60	HD battery ($15.80 w/325 hp, 350 hp or 375 hp engine)	9,738	8.45
U15	Speed warning indicator	2,111	11.60
U16	Tachometer gauge	1,410	52.70
U17	Special instrumentation	29,524	94.80
U35	Electric clock	20,330	15.80
U46	Light monitoring system	1,450	26.35
U57	Stereo tape system	6,239	133.80
U63	Push-button AM radio	206,598	61.10
U69	Push-button AM/FM radio	8,271	133.80
U73	Manual rear antenna (NA w/AM/FM radio or spoiler equipment)	16,394	9.50
U79	Push-button AM/FM stereo radio	2,359	239.10

U80	Rear seat speaker	26,862	13.20
VE3	Special front bumper	12,650	42.15
V01	HD radiator	3,802	14.75
V31	Front bumper guard	12,657	12.65
V32	Rear bumper guard	12,369	12.65
V75	Liquid tire chain	188	23.20
ZJ7	Rally wheels	48,735	35.85
ZJ9	Auxiliary Lighting Group	15,768	13.70
ZK3	Custom deluxe belts	18,760	12.15
ZL2	Special ducted hood (available only when Z27 or Z28 is ordered)	10,026	79.00
Z11	Indy sport convertible accents	3,675	36.90
Z21	Style Trim Group	102,740	47.40
Z22	Rally Sport Package	37,773	131.65
Z23	Special interior	66,469	17.95
Z27	Camaro SS equipment	34,932	295.95
Z28	Special Performance Package	20,302	458.15
Z87	Custom interior	39,875	110.60
—	Optional axle ratios	—	2.15
—	Two-tone paint	5,909	31.60
—	Vinyl roof	100,602	84.30

Facts

The 1969 Camaro was extensively facelifted, giving it a more aggressive look.

The SS identification included emblems on the grille, taillight panel, both front fenders behind the wheelwells and steering wheel. All Super Sport models came with a hood that included the eight simulated carburetor stacks. Engine numeral identification was located in front of the side marker lights on the front fenders. Different Super Sport side striping was used. Resembling a hockey stick, the stripes began at the front of the fender and followed the top of the fender and body to just before the door handles. The Super Sport models also got simulated rear fender louvers.

The Rally Sport option was available in conjunction with the SS Package. The headlamp doors were louvered for 1969, in case the doors malfunctioned. Rear taillights were nonsegmented, with the back-up lamps relocated below the bumper. Other Rally Sport features included headlamp washers, wheelwell moldings and black rocker panels on cars painted in lighter colors. As in previous years, SS identification superseded RS identification.

The VE3 optional front bumper, more commonly known as the Endura bumper, was painted to match the car's body color.

Standard Super Sport engine was the 300 hp 350 ci V-8, which this year had four-bolt mains. Optional were the 325 hp, 350 hp and 375 hp versions of the 396 ci big-block V-8. The L89 version with aluminum cylinder heads was also available. During the model year, the 396 engine's displacement increased to 402 ci. The engine was still marketed as a 396 and not a 402.

A set of F70x14 RWL tires on 14x7 in. steel rims was standard.

Power front disc brakes were included with the Super Sport option.

The Turbo Hydra-matic was available with all Super Sport engines. Other mechanical improvements included the availability of fast-ratio power steering, which was included when power steering was ordered. Even quicker steering, RPO 44, was included when power steering was ordered with Super Sport equipped Camaros.

All four-speed-manual-equipped cars came with a Hurst shifter.

Four-wheel disc brakes were a rare option, available only on Z-28 or Super Sport Camaros. The regular front disc brakes used a new single piston caliper in 1969.

The year 1969 was the first in which the alternator was mounted on the right (passenger's) side. In 1968, small-block engines had it mounted on the left (driver's) side of the engine.

Air conditioning was not available with the 375 hp 396 ci engine. The only manual transmission available with the 375 hp engine was the M22.

The cowl induction hood was optionally available on the Super Sport.

Front and rear spoilers were also optional.

The chambered exhaust system was initially available, but it was withdrawn because it was too loud.

A total 3,675 Indy pace car convertible replicas were built. All were Super Sport, Rally Sport models with a Dover White exterior paint with Orange Houndstooth interiors. The cowl induction hood was used with Z-28 type hood and rear deck stripes, along with rally wheels and bright exhaust tips.

This was the last year for the Super Sport convertible.

The 1969 Camaro SS 350 Sport Coupe.

Chapter 23

1970 Camaro SS

Production

8 cyl
2 dr coupe   112,323

V-8
2 dr coupe, L48 350 ci   10,012

2 dr coupe, L34 396 ci   1,864
2 dr coupe, L78 396 ci   600
Total   12,476

Serial numbers

Description
124370L100001
12437 — Model number (12437-2 dr coupe)
0 — Last digit of model year (1970)
L — Assembly plant (L-Los Angeles, N-Norwood)
100001 — Consecutive sequence number

Location
     On plate attached to driver's side of dash panel, visible through the windshield.

Engine and transmission suffix codes
CNJ — 350 ci V-8 4 bbl 300 hp, manual
CNK — 350 ci V-8 4 bbl 300 hp, Powerglide automatic
CRE — 350 ci V-8 4 bbl 300 hp, Turbo Hydra-matic automatic
    TH350
CJF — 396 ci V-8 4 bbl 350 hp, manual
CJI — 396 ci V-8 4 bbl 350 hp, Turbo Hydra-matic automatic
CJH — 396 ci V-8 4 bbl 375 hp, manual
CJK — 396 ci V-8 4 bbl 375 hp, Turbo Hydra-matic automatic

Carburetors
350 ci — 7040203
350 ci w/automatic — 7040202
396 ci — 7040205
396 ci w/automatic — 7040204
396 ci 375 hp w/EEC — 3957477 (Holley R-4557A)
396 ci 375 hp w/EEC — 3967479 (Holley R-4491A)
396 ci 375 hp w/EEC & automatic — 3969898 (Holley R-4492A)
396 ci 375 hp w/EEC & automatic — 3969894 (Holley R-4556A)

Distributors
350 ci — 1111997
396 ci — 11112000
396 ci 375 hp — 1112000

Exterior color codes

Classic White	10	Forest Green	48
Cortez Silver	14	Daytona Yellow	51
Shadow Gray	17	Camaro Gold	53
Astro Blue	25	Autumn Gold	58
Mulsanne Blue	26	Desert Sand	63
Citrus Green	43	Hugger Orange	65
Green Mist	45	Classic Copper	67
		Cranberry Red	75

Interior trim codes

Color	Std	Custom cloth	Custom vinyl
Black	711	725	712
Blue	715	—	716
Black/Blue	—	714	—
Dark Green	723	—	724
Black/Dark Green	—	720	—
Saddle	726	—	727
Sandlewood	710	—	730
Black/White	—	713	—

Vinyl top color codes

White	AA	Dark Green	GG
Black	BB	Dark Gold	HH
Dark Blue	CC		

Options

12487 Sport coupe $2,839.00

Option number	Description	Quantity	Retail price
AK1	Custom deluxe belts	13,218	$ 12.15
AS4	Rear shoulder belts	89	26.35
A01	Tinted glass (all windows)	71,363	37.95
B37	Color-keyed floor mats	23,708	11.60
B93	Door edge guards	35,577	5.30
C08	Vinyl roof cover	43,221	89.55
C50	Forced air rear window defroster	8,814	26.35
C60	AC	38,565	380.25
D34	Vanity Visor mirro	7,423	3.20
D35	Sport exterior mirrors (incl RH mirror & remote control LH mirror)	31,726	26.35
D55	Console	—	59.00
D80	Rear deck spoiler	—	32.65
F41	Special performance front & rear suspension	—	30.55
G80	Positraction axle	19,752	44.25
L34	350 hp Turbo-Jet 396 ci V–8 engine (available only w/Z27)	1,864	152.75

Code	Description	Qty	Price
L78	375 hp Turbo-Jet 396 ci V–8 engine (incl special performance suspension)	600	385.50
M20	4 speed wide-ratio transmission	12,191	205.95
M40	Turbo Hydra-matic automatic transmission	71,832	290.40
NA9	EEC (Calif only)	15,862	36.90
N33	Comfortilt steering wheel	6,735	45.30
N40	Variable-ratio power steering	92,640	105.35
PL4	F70x14/B bias-belted-ply white letter tires (incl 14x7 in. wheels; incl w/Z27)	20,783	65.35
PX6	F78–14/B bias-belted-ply white stripe tires	12,893	43.30
PY4	F70x14B bias-belted-ply white stripe tires (incl 14x7 in. wheels)	15,776	65.70
P01	Brightmetal wheel covers	73,292	26.35
P02	Special wheel covers	3,532	79.00
T60	HD battery	5,518	15.80
U14	Special instrumentation	17,842	84.30
U35	Electric clock (incl w/U14)	15,533	15.80
U63	Push-button AM radio	110,779	61.10
U69	Push-button AM/FM radio	8,586	133.80
U80	Rear seat speaker	20,583	14.75
VF3	Deluxe front & rear bumpers	1,605	36.90
V01	HD radiator	1,509	14.75
YD1	Special ratio axle for trailering	132	12.65
ZJ7	Rally wheels (available only w/F70–14 tires)	15,197	42.15
ZJ9	Auxiliary lighting	8,307	13.70
ZQ9	Performance ratio axle (available only w/375 hp engine & Positraction rear axle)	3,161	12.65
Z21	Style trim	43,344	52.70
Z22	Rally sport equipment	27,136	168.55
Z23	Interior Accent Group	36,550	21.10
Z27	Camaro SS equipment	12,476	289.65
Z28	Special Performance Package	8,733	572.95
Z87	Custom interior	21,059	115.90

Facts

The Camaro was totally restyled for 1970. This second-generation body design continued until 1981.

The Super Sport consisted of a blacked-out grille; rear hood molding; dual sport mirrors; and SS emblems on the grille, rear deck and both front fenders behind the wheelwells. The fender emblems also included engine size numbering. A set of 14x7 in. wheels with F70x14 RWL tires, the F41 Suspension Package and power front disc brakes were included. Power variable-ratio steering was optional.

The rear deck panel was painted black with 396 ci equipped Camaros.

In the interior, an SS emblem was used on the steering wheel.

The smallest engine available was the 300 hp 350 ci small-block. Optional were the 350 hp and 375 hp versions of the 396 ci, actually displacing 402 ci. All Super Sport Camaros got bright dual-exhaust tips. Super Sport Camaros could be equipped with a four-speed manual or the Turbo Hydra-matic automatic.

The RS Option Package was again available on Super Sport equipped cars, but without hidden headlights. With the Rally Sport, bumperettes under each headlight replaced the stock full-length bumper. The stock turn lamp units were then replaced by round lamps located next to the headlights. The grille was extended and framed in body-colored urethane to give the Camaro a distinctively European look. In the rear, the taillights got bright accents.

The conventional radio antenna was replaced by a wire antenna imbedded in the windshield.

Black exterior paint was not available.

Chapter 24

1971 Camaro SS

Production

8 cyl
2 dr sport coupe   103,452

V-8
2 dr coupe, L48 350 ci   6,844
2 dr coupe, LS3 396 ci   1,533
  Total   8,377

Serial numbers

Description
124371L100001
12437 — Model number (12437-2 dr coupe)
1 — Last digit of model year (1971)
L — Assembly plant (L-Los Angeles, N-Norwood)
100001 — Consecutive sequence number

Location
On plate attached to driver's side of dash, visible through the windshield.

Engine and transmission suffix codes
CGK, CJG — 350 ci V-8 4 bbl 270 hp, manual
CGL, CJD — 350 ci V-8 4 bbl 270 hp, Turbo Hydra-matic automatic
  TH350
CLC — 402 ci V-8 4 bbl 300 hp, manual
CLD — 402 ci V-8 4 bbl 300 hp, Turbo Hydra-matic automatic

Carburetors
350 ci — 7041203
350 ci w/automatic — 7041202
402 ci — 7041201
402 ci w/automatic — 7041200

Distributors
350 ci — 1112044
350 ci w/automatic — 1112045
402 ci — 1112057

Exterior color codes

Antique White	11	Sunflower	52
Nevada Silver	13	Placer Gold	53
Tuxedo Black	19	Sandalwood	61
Ascot Blue	24	Burnt Orange	62
Mulsanne Blue	26	Classic Copper	67
Cottonwood Green	42	Cranberry Red	75
Lime Green	43	Rosewood	78
Antique Green	49		

Interior trim codes

Color	Std vinyl	Custom cloth
Black	775	785
Blue	776	786
Dark Jade	778	787
Dark Saddle	779	792
Sandalwood	777	—
Black/White	—	789

Vinyl top color codes

White	AA
Black	BB
Blue	CC
Brown	FF
Green	GG

Options

12487 Sport coupe $2,848.00

Option number	Description	Quantity	Retail price
AK1	Custom deluxe belts	16,922	$ 15.30
AN6	2 position adjustable seatback	—	19.00
AS4	Rear shoulder belts	99	26.35
A01	Tinted glass (all windows)	67,250	40.05
B37	Color-keyed floor mats	22,576	6.35
B93	Door edge guards	33,124	6.35
C08	Vinyl roof	38,329	89.55
C50	Forced air rear window defroster	8,794	31.60
C60	AC	42,537	402.35
D34	Vanity Visor mirror	5,522	3.20
D35	Remote control LH mirror (incl w/Z27)	40,684	15.80
D55	Console	72,656	59.00
D80	Front & rear spoiler	6,489	79.00
F41	Special performance suspension (incl w/300 hp Z27)	10,975	30.55
G80	Positraction axle	11,753	44.25
J50	Power brakes (incl w/Z27)	41,630	47.30
LS3	300 hp Turbo-Jet 396 ci V-8 engine	1,533	99.05
M20	4 speed wide-ratio transmission	7,603	205.95
M21	4 speed close-ratio transmission	1,721	205.95
M40	Turbo Hydra-matic automatic transmission	77,541	306.25
NK2	Custom steering wheel (black; NA w/N33)	621	15.80
NK4	Sport steering wheel (4 spoke; black)	6,216	15.80
N33	Comfortilt steering wheel	8,374	45.30
N40	Variable-ratio power steering	93,163	110.60

PY4	F70-14B bias-belted-ply white stripe tries	24,579	NC
P01	Brightmetal wheel covers	55,363	26.35
P02	Special wheel covers	1,809	84.30
T60	HD battery	5,168	15.80
U14	Special instrumentation	12,174	84.30
U35	Electric clock (incl w/U14)	10,338	16.90
U63	Push-button AM radio	95,776	66.40
U69	Push-button AM/FM radio	13,310	139.05
U80	Rear seat speaker	20,018	15.80
VF3	Deluxe front & rear bumpers	1,309	36.90
V01	HD radiator	1,594	14.75
ZJ7	Rally wheels (incl special 14x7 in. wheels, hubcaps & trim rings)	34,604	45.30
ZJ9	Auxiliary lighting	6,323	18.45
Z21	Style trim	38,161	57.95
Z22	Rally sport equipment	18,404	179.05
Z27	Camaro SS equipment	8,377	313.90
Z28	Special Performance Package	4,862	786.75
Z87	Custom interior	11,643	115.90

Facts

Detail changes characterized the 1972 Camaro. Some of these were new emblems, paint colors, wheel covers and a new high-back bucket seat with seatback adjustment. The D80 Spoiler Package included front and rear spoilers and was available on the Super Sport.

The Z27 Super Sport option was basically the same. It included a blacked-out grille with SS emblem, SS emblems with engine size numerals on the front fenders, dual sport mirrors, dual bright exhaust tips, the F41 Suspension Package, power front disc brakes, 14x7 in. steel wheels with F70x14 RWL tires, the hideaway wipers and an SS emblem on the steering wheel. All 396 ci powered Camaros also got the black-painted rear deck lid.

Engine availability was limited to just two, the 270 hp (210 hp net) 350 ci small-block or the 300 hp (260 hp net) 396 ci (402 ci actual displacement) big-block. As before, either engine was available with a four-speed manual or the Turbo Hydra-matic automatic transmission. Both engines still came with chrome air cleaners and chrome valve covers (finned aluminum with the 350 ci). The lower horsepower ratings were due to tuning and a reduced compression ratio as well as measurements produced by the more realistic SAE net method.

The RS Package was again available in conjunction with the Super Sport.

The 1971 Camaro SS Coupe.

Chapter 25

1972 Camaro SS

Production

8 cyl
2 dr coupe   63,830

V-8
2 dr coupe, L48 350 ci   5,592
2 dr coupe, LS3 402 ci   970
   Total   6,562

Serial numbers

Description
1F87K2N100001
1 — Chevrolet
F — Camaro body series
87 — Body style (2 dr coupe)
K — Engine code
2 — Last digit of model year (1972)
N — Assembly plant (N-Norwood)
100001 — Consecutive sequence number

Location
   On plate attached to driver's side of dash, visible through the windshield.

Engine codes
K — 350 ci 4 bbl V-8 200 hp
U — 402 ci 4 bbl V-8 240 hp

Engine and transmission suffix codes
CKK — 350 ci V-8 4 bbl 200 hp, manual
CDG — 350 ci V-8 4 bbl 200 hp, manual w/NB2
CKD — 350 ci V-8 4 bbl 200 hp, Turbo Hydra-matic automatic
   TH350
CDD — 350 ci V-8 4 bbl 200 hp, Turbo Hydra-matic automatic
   w/NB2
CLA — 402 ci V-8 4 bbl 240 hp, manual
CTA — 402 ci V-8 4 bbl 240 hp, manual w/AIR
CLB — 402 ci V-8 4 bbl 240 hp, Turbo Hydra-matic automatic
CTB — 402 ci V-8 4 bbl 240 hp, Turbo Hydra-matic automatic

Carburetors
350 ci — 7042203
350 ci w/automatic — 7042202

350 ci w/NB2 — 7042903
350 ci w/NB2, w/automatic — 7042902
402 ci — 7042201
402 ci w/automatic — 7042200

Distributors

350 ci — 1112095
350 ci w/automatic —
1112049
402 ci — 1112057

350 ci w/automatic — 1112154
402 ci — 1112162

Exterior color codes

Antique White	11
Pewter Silver	14
Ascot Blue	24
Mulsanne Blue	26
Spring Green	36
Gulf Green	43
Sequoia Green	48
Covert Tan	50
Placer Gold	53
Cream Yellow	56
Golden Brown	57
Mohave Gold	63
Orange Flame	65
Midnight Bronze	68
Cranberry Red	75

Interior trim codes

Color	Std vinyl	Custom cloth
Black	775	785
Dark Blue	776	786
Dark Green	777	787
Light Covert	779	788
Medium Tan	778	—
White	780	—

Vinyl top color codes

White	AA
Black	BB
Medium Tan	FF
Medium Green	GG
Light Covert	TT

Options

12487 Sport coupe $2,819.70

Option number	Description	Quantity	Retail price
AK1	Custom deluxe seatbelts & shoulder belts	8,475	$ 14.50
AN6	2 position adjustable seatback	2,087	18.00
A01	Tinted glass (all windows)	44,155	39.00
B37	Color-keyed floor mats	15,725	12.00
B93	Door edge guards	21,452	6.00
C08	Vinyl roof cover	23,918	87.00
C50	Forced air rear window defroster	7,018	31.00
C60	AC	31,738	397.00
D34	Vanity Visor mirror	3,931	3.00
D55	Rear seat & ashtray console	49,845	57.00
D80	Front & rear spoiler	5,954	77.00
F41	Special performance suspension & rear shock absorbers (V-8 models w/F70x14 tires only)	7,133	30.00
G80	Positraction axle	7,643	45.00

LS3	240 hp Turbo-Jet 396 ci V-8 engine (available only w/Z27)	970	96.00
M20	4 speed wide-ratio transmission	4,127	200.00
M21	4 speed close-ratio transmission	942	200.00
M40	Turbo Hydra-matic automatic transmission	7,302	297.00
NK4	Sport steering wheel	5,758	15.00
N33	Comfortilt steering wheel	3,706	44.00
N40	Variable-ratio power steering	59,854	130.00
PY4	F70-14B bias-belted-ply white stripe tires (incl 14x7 in. wheels)	16,581	NC
P01	Brightmetal wheel covers	27,708	26.00
P02	Custom wheel covers	824	82.00
T60	HD battery	3,448	15.00
U14	Special instrumentation	8,608	82.00
U35	Electric clock (incl w/U14)	7,403	16.00
U63	Push-button AM radio	10,404	135.00
U80	Rear seat speaker	15,889	15.00
VF3	Deluxe front & rear bumpers	2,449	36.00
V01	HD radiator	3,057	14.00
YD1	Special ratio axle for trailering	165	12.00
YF5	Calif emission test (NA w/240 hp engines)	8,124	15.00
ZJ7	Rally wheels (incl special 14x7 in. wheels, hubcaps & trim rings)	27,804	44.00
ZJ9	Auxiliary lighting	5,309	17.50
Z21	Style trim	22,477	56.00
Z22	Rally sport equipment	11,364	118.00
Z23	Interior Accent Group	18,064	21.00
Z27	Camaro SS equipment	6,562	306.35
Z28	Special Performance Package	2,575	769.15
Z87	Custom interior	6,462	113.00

Facts

The year 1972 was the last for the SS Package on the Camaro and also the last for the big-block 402 ci engine, which was marketed as the 396.

From 1972, Chevrolet revised the VIN system; the car's VIN now included a letter code indicating which engine the car was equipped with.

All 1972 Camaros were built at the Norwood plant.

Most noticeable changes on the 1972 Camaro were the use of a larger-mesh standard grille and redesigned inner door panels. In the interior was a new three-point harness and seatbelt combination and a new shifter with pushdown reverse lockout for four-speed-manual-equipped cars.

The SS Package was unchanged from 1971; however, factory literature mentioned that Super Sport Camaros got heavy-duty engine mounts and starter. The RS Package was unchanged as well.

The standard Super Sport 350 ci V-8 was downgraded to produce just 200 hp. The optional 402 was rated at 240 hp. Ratings were based on the SAE net rating system.

The 1972 Camaro SS Coupe.

Chapter 26

1970 Monte Carlo SS 454

Production
13857 2 dr coupe, 8 cyl   3,823

Serial numbers

Description
138570A100001
13857 — Model number (13857-2 dr coupe)
0 — Last digit of model year (1970)
A — Assembly plant (A-Lakewood, B-Baltimore, F-Flint, K-Leeds, L-Van Nuys, 1-Oshawa)
100001 — Consecutive sequence number

Location
   On plate attached to driver's side of dash, visible through the windshield.

Engine and transmission suffix codes
CGW — 454 ci V-8 4 bbl 360 hp, 4 speed manual
CGT — 454 ci V-8 4 bbl 360 hp, Turbo Hydra-matic automatic

Carburetor
454 ci — 7040501
454 ci w/automatic — 7040500

Distributor
454 ci — 1108418
454 ci w/automatic — 1108430

Exterior color codes
Classic White	10
Cortez Silver	14
Shadow Gray	17
Tuxedo Black	19
Astro Blue	25
Fathom Blue	28
Misty Turquoise	34
Green Mist	45
Forest Green	48
Gobi Beige	50
Champagne Gold	55
Autumn Gold	58
Desert Sand	63
Cranberry Red	75

Two-tone color codes*
Astro Blue/	
Classic White	25/10
Misty Turquoise/	
Classic White	34/10
Fathom Blue/Astro Blue	28/25
Astro Blue/Fathom Blue	25/28
Champagne Gold/	
Classic White	55/10
Autumn Gold/	
Classic White	58/10
Desert Sand/	
Classic White	63/10

*Lower/upper.

Interior trim codes

Color	Cloth bench seats	Vinyl bench seats	Cloth bucket seats	Vinyl bucket seats
Black	748	—	749	757
Blue	758	—	—	—
Dark Blue	767	—	—	—
Gold	774	—	—	—
Dark Green	780	—	—	784
Saddle	—	—	—	789
Sandalwood	792	—	—	—

Vinyl top color codes

White	AA
Black	BB
Dark Blue	CC
Dark Green	GG
Dark Gold	HH

Options*
13857 Coupe $3,123.00

Option number	Description	Retail price
AK1	Seatbelts & shoulder belts	$ 12.15–13.70
AQ2	Automatic seatbelt latch	23.70
AS4	2 rear shoulder belts (requires AK1)	26.35
AU3	Power door lock system	44.80
A01	Soft Ray tinted glass (all windows)	42.15
A31	Power windows	105.35
A41	4 way electric power seat (w/bench seat)	73.75
A46	4 way electric power seat (w/bucket seats)	73.75
A51	Strato-Bucket seats	121.15
A90	Power trunk opener	14.75
B37	2 front & 2 rear color-keyed floor mats	11.60
B85	Belt moldings	19.00
B93	Door edge guards	15.80
CD3	Fingertip windshield wiper control	19.00
C50	Forced air rear window defroster	26.35
C60	Four Season AC	376.00
D33	Remote control outside rearview mirror	10.55
D34	Vanity Visor mirror	3.20
D55	Console	53.75
G67	Rear shock absorber	89.55
G80	Positraction rear axle	42.15
K05	Engine block heater	10.55
K30	Cruise-Master speed control	57.95
K85	63 amp Delcotron generator	5.30–26.35

LF6	265 hp Turbo-Fire 400 V–8 engine (regular fuel)	63.20
LS3	330 hp Turbo-Jet 400 V–8 engine	141.15
L48	300 hp Turbo-Jet 350 V–8 engine	47.40
M20	4 speed wide-range transmission	184.80
M35	Powerglide transmission	174.25
M40	Turbo Hydra-matic transmission	184.80–221.80
NA9	EEC	36.90
NK1	Cushioned-rim steering wheel	34.80
N10	Dual exhaust	30.55
N33	Comfortilt steering wheel	57.95
N40	Power steering	105.35
PA3	Color-keyed wheel covers	15.80
PH1	15x7 in. JK wheels	10.55
PU8	G78–15/B belted white stripe tubeless tires	30.20
P02	Special wheel covers	73.75–84.30
P90	G70–15/B belted white stripe tubeless tires	46.95
T58	Rear fender skirts	31.60
T60	HD 30 amp-hr battery	15.80
UM1	Stereo tape system (w/AM radio)	194.85
UM2	Stereo tape system (w/AM/FM stereo radio)	372.85
U14	Special instrumentation	68.50
U46	Light monitoring system	26.35
U63	Push-button AM radio	61.10
U69	Push-button AM/FM radio	133.80
U79	Push-button AM/FM stereo radio	239.10
U80	Rear seat speaker	13.20
V01	HD radiator	14.75
V31	Front bumper guards	15.80
V32	Rear bumper guards	15.80
ZJ7	Rally wheels	31.60
ZJ9	Auxiliary lighting	25.30
ZP5	Appearance Guard Group	50.65
ZQ2	Operating Convenience Group	36.90
Z20	Monte Carlo SS equipment	420.25

*Quantities are included with 1970 Chevelle options.

Facts

The Monte Carlo was Chevrolet's version of the new G-body platform that was first seen in 1969 with the Pontiac Grand Prix. It was a stretched version of the A-platform that was used for the Chevelle intermediate, thus making most, if not all, mechanical components interchangeable with those for the Chevelle.

Exterior identification on the Monte Carlo was limited to SS 454 rocker panel emblems behind the front wheelwells. All Super Sport Monte Carlos came with twin chrome exhaust pipe extensions.

Only one engine was available, the LS5 454 ci big-block rated at 360 hp. It featured the small oval-port cylinder heads, cast-iron intake manifold with a Quadrajet carburetor and a hydraulic camshaft. The Turbo Hydra-matic transmission and heavy-duty battery were mandatory options with the 454.

A set of 15x7 wheels with G70x15 whitewall tires was standard. Heavy-duty front and rear springs and shocks along with front and rear stabilizer bars were also part of the Monte Carlo SS.

An interesting feature was the Superlift rear shock absorbers with automatic level control. Working off engine vacuum was a small air compressor that maintained rear ride height.

Cruise control was not available with the Monte Carlo SS.

Monte Carlos came with the new wire antenna imbedded in the windshield, rather than a conventional unit.

The 1970 Monte Carlo SS 454 Coupe.

Chapter 27

1971 Monte Carlo SS 454

Production
13857 2 dr coupe, 8 cyl   1,919

Serial numbers

Description
138571B100001
13857 — Model number (13857-2 dr coupe)
1 — Last digit of model year (1971)
B — Assembly plant (B-Baltimore, K-Leeds, L-Van Nuys,
    R-Arlington, 1-Oshawa)
100001 — Consecutive sequence number

Location
On plate attached to driver's side of dash, visible through the windshield.

Engine and transmission suffix codes
CPA — 454 ci V-8 4 bbl 365 hp, 4 speed manual
CPO — 454 ci V-8 4 bbl 365 hp, Turbo Hydra-matic automatic

Carburetor
454 ci — 7041201
454 ci w/automatic — 7041200

Distributor
454 ci — 1112052

Exterior color codes

Antique White	11
Nevada Silver	13
Tuxedo Black	19
Ascot Blue	24
Mulsanne Blue	26
Cottonwood Green	42
Lime Green	43
Antique Green	49
Sunflower	52
Placer Gold	53
Sandalwood	61
Burnt Orange	62
Classic Copper	67
Cranberry Red	75
Rosewood	78

Two-tone color codes*

Mulsanne Blue/	
Antique White	25/11
Placer Gold/	
Antique White	53/11
Antique Green/	
Antique White	49/11
Lime Green/	
Antique White	43/11
Burnt Orange/	
Antique White	62/11
Sandalwood/	
Antique White	61/11
*Lower/upper.	

Interior trim codes

Color	Cloth bench seats	Vinyl bench seats	Cloth bucket seats	Vinly bucket seats
Black	708	—	707	710
Dark Blue	728	—	727	—
Dark Jade	734	—	733	729
Saddle	—	—	—	723
Sandalwood	717	—	716	—

Vinyl top color codes

White		AA	Brown	FF
Black		BB	Green	GG
Blue		CC		

Options

13857 Coupe $3,304.00

Option number	Description	Retail price
AA	White vinyl roof cover (incl bright outline moldings)	$126.40
BB	Black vinyl roof cover (incl bright outline moldings)	126.40
CC	Dark Blue vinyl roof cover (incl bright outline moldings)	126.40
FF	Dark Brown vinyl roof cover (incl bright outline moldings)	126.40
GG	Dark Green vinyl roof cover (incl bright outline moldings)	126.40
AK1	Custom deluxe seatbelts & shoulder belts	15.30–16.90
AS4	Custom deluxe shoulder belts	26.35
AU3	Power door lock system	46.35
A01	Soft Ray tinted glass (all windows)	46.35
A31	Power windows	127.45
A41	4 way electric power seat (w/bench seat)	79.00
A46	4 way electric power seat (w/bucket seats; driver's seat only)	79.00
A90	Power trunk opener	14.75
B37	2 front & 2 rear color-keyed floor mats	12.65
B85	Belt moldings	19.00
B93	Door edge guards	6.35
C50	Forced air rear window defroster	31.60
C60	Four Season AC	407.60
D33	Remote control outside LH rearview mirror	12.65
D34	Vanity Visor mirror	3.20
D55	Console	59.00
G67	Rear shock absorber	89.55

Code	Description	Price
G80	Positraction rear axle	46.35
K30	Cruise-Master speed control	63.20
K85	63 amp Delcotron generator	5.30–26.35
LS3	300 hp Turbo-Jet 400 V–8 engine	146.40
L48	270 hp Turbo-Fire 350 V–8 engine	47.40
M20	4 speed wide-range transmission (w/270 hp or 300 hp engine only)	195.40
M35	Powerglide transmission (w/std engine only)	190.10
M40	Turbo Hydra-matic transmission	216.50–237.60
NK2	Custom steering wheel (black)	15.80
NK4	Sport steering wheel (4 spoke; black)	15.80
N33	Comfortilt steering wheel (requires opt transmission)	45.30
N40	Power steering	115.90
PA3	Deluxe wheel covers	15.80
PH1	15x7 in. JK wheels (incl w/Monte Carlo SS, rally wheels or custom wheel covers)	10.55
PU8	G78-15/B belted white stripe tubeless tires	32.30
P02	Custom wheel covers (incl 15x7 in. wheels)	80.05
P90	G70-15/B belted white stripe tubeless tires	49.35
T58	Rear fender skirts (NA w/15x7 in. wheels)	31.60
T60	HD 80 amp-hr battery	15.80
UM1	Stereo tape system (w/AM radio)	200.15
UM2	Stereo tape system (w/AM/FM stereo radio)	372.85
U14	Special instrumentation	68.50
U63	Push-button AM radio	66.40
U69	Push-button AM/FM radio	139.05
U79	Push-button AM/FM stereo radio	239.10
U80	Rear seat speaker (NA when stereo is ordered)	15.80
V01	HD radiator	21.10
V30	Bumper guards	15.80–31.60
YD1	Trailering axle ratio	12.65
ZJ7	Rally wheels (incl w/Monte Carlo SS; incl special 15x7 in. wheels, hubcaps & trim rings)	41.10
ZJ9	Auxiliary lighting (incl ashtray, courtesy, luggage compartment, mirror, map & underhood lights)	21.10
ZP5	Vanity Visor mirror	38.00–53.80

| ZQ2 | Remote control outside LH rearview mirror | 12.65– 44.25 |
| Z20 | Monte Carlo SS equipment | 484.50 |

Quantities are included with 1971 Chevelle options.

Facts

The Monte Carlo was mildly restyled for 1971. Most noticeable changes were the finer-grid front grille, wider-spaced headlights, larger headlight bezels and rectangular turn signal lights in the front bumper.

Super Sport 454 Monte Carlos came with an SS emblem on the black rear trunk panel in addition to the rocker panel emblems. Replacing the regular 15x7 steel wheels of 1970 were 15x7 in. rally wheels. Power front disc brakes were standard as well.

The LS5 454 ci engine was rated at 365 even though compression dropped to 8.5:1. When measured in the SAE net method, which General Motors introduced in 1971, it was rated at 285 hp. The Turbo Hydra-matic automatic was standard with the 454. A four-speed manual may have been available.

Production records indicate that 1,919 RPO Z20 Monte Carlo SSs were built. However, a discrepancy arises when this figure is compared with the 1971 Chevrolet Engine Production by Cubic Inch Displacement report. It lists 1,772 of the 454 equipped Monte Carlos. One or the other figure may be incorrect, or perhaps both are correct and 147 Super Sport models got another engine, such as the LS3 402 ci V-8. No LS6 454s were installed in any 1971 production cars.

The automatic level control was part of the 1971 SS Package.

Appendix

Engine identification

Every Chevrolet engine has a number stamped on it to identify it and connect it with the car it is installed in. This number consists of the engine code and part of the car's VIN.

The engine code consists of the engine plant code, four digits indicating the date and month the engine was assembled, and the engine suffix code. Engine plant codes are as follows:

F — Flint Motor
H — Hydramatic
K — GM of Canada
M — GM of Mexico
S — Saginaw
T — Tonawanda
V — Flint Engine

The four digits indicating date and month are broken down as follows: the first two indicate month and the other two indicate date. For example, 1109 decodes to November 9.

The suffix code indicates engine family, usage and, most of the time, type of transmission. Before 1970, this was a two-letter code.

From 1970, this was a three-letter code. All suffix codes are listed in the chapters of this book.

On V–8 engines, the engine code is stamped on a pad on the right (passenger's) side of the engine block just where the cylinder head and block meet. Also on the pad may be found the last eight digits of the car's VIN. This ties in the installed engine with the chassis. From 1969, you may find that these numbers are also stamped on a pad near the oil filter, or they may just be stamped on the filter pad alone. However, most engines will have the number stamped on the front of the block as indicated.

As an example, the identification code T0422CRV decodes as follows:

T — Tonawanda

0422 — April 22

CRV — 454 ci 450 hp with manual transmission

Tonawanda is the only plant where the big-block 396 ci, 402 ci, 427 ci and 454 ci engines were built.

Engine casting date codes

Although it is beyond the scope of this book to list all engine part casting numbers, it is useful to be able to decode the date a part was cast. Most parts used on an engine should predate the assembly date code of the engine and should be within thirty days of engine assembly. Exceptions do exist, such as parts cast for use at a later date or in a later model year.

Engine casting date codes consist of three or four digits, beginning with a letter A through L to represent a month January through December. The next digit or pair of digits is a number that stands for the date of the month, and the last digit is a number that stands for the last number of the model year. For example, the code B228 stands for February 22, 1968, and the code H71 stands for August 7, 1971.

The date code is located on the right (passenger's) rear side of the block on six-cylinder and small-block V–8s. On big-block V–8s, it is stamped on the right side of the block in front of the starter.

In much the same way, subsidiary parts, such as manifolds, carry a similar casting date.

Transmission identification

After the engine, the transmission is the next major component that should be checked. Until 1967, transmissions carried a source serial number, which consisted of a plant prefix letter that stood for the transmission plant, the production date and a shift suffix indicating whether the unit was built during the day or night shift.

The following are plant prefix letters:

Prefix	Plant	Transmission type
A	Cleveland	Manual Powerglide
B	Cleveland	Turbo Hydra-matic
C	Cleveland	Powerglide
CA	Hydramatic	Turbo Hydra-matic

D	Saginaw	Overdrive
E	McKinnon	Powerglide
H	Muncie	3 speed
K	McKinnon	3 speed
L	GM of Canada	Turbo Hydra-matic
M	Muncie	3 speed & overdrive
N	Muncie	4 speed
O	Saginaw	Overdrive
P	Warner Gear	3 & 4 speed
P	Muncie	4 speed
R	Muncie	4 speed
R	Saginaw	4 speed
S	Muncie	3 speed
S	Saginaw	3 speed
T	Toledo	Powerglide
X	Cleveland	Turbo Hydra-matic
Y	Toledo	Turbo Hydra-matic

For example, the number M605 decodes to a Muncie three-speed built on June 5, and the number C213N decodes to a Cleveland Powerglide built on February 13 during the night shift.

For 1967, the transmission source serial number was changed to include the model year. Rather than using a number to indicate the month, it used a letter, as follows:

A — January	E — May	P — September
B — February	H — June	R — October
C — March	K — July	S — November
D — April	M — August	T — December

For example, the number N7B08 decodes to a four-speed Muncie for model year 1967, built on February 8.

For example, 67A455 decodes to model year 1967, an engine or vehicle designated by the letter *A* and the build date March 31, 1967.

Effective October 21, 1968, an additional letter was added to the plant prefix number to facilitate the identification of Muncie transmission ratios. The additional letter codes were as follows:

Muncie three-speed manual transmission

Suffix	1st gear ratio
A	3.03:1
B	2.42:1

Muncie four-speed manual transmission

Suffix	1st gear ratio
A	2.52:1 wide range
B	2.20:1 close range
C	2.20:1 Rock Crusher

In addition, beginning in 1968, the last eight numbers of the car's VIN were stamped on the transmission. Automatic transmissions were stamped on the left side near the top forward portion of the housing. Manual transmissions were stamped on the case.

The location of the car's VIN varied in subsequent years, but it was stamped on the transmission.

Cowl tags

An important way to identify a Chevrolet is by the cowl tag. Cowl tags were thin sheet metal tags with stamped numbers and letters that were riveted on the left side of the cowl in the engine compartment.

Several types were used. The first, used until 1964, contained four pieces of information: style number (model year, series and type), body number, trim number and paint number. The consecutive sequence number was not included on the plate.

Between 1964 and 1967, a different, more informative plate was used. Line 1 contained the time built code. This consisted of two numbers followed by a letter. The numbers ranged from 01 to 12 and represented the months of the year. The letter (A, B, C, D or E) represented the week of production (first, second, third, fourth or fifth). For example, 03C decodes to the third week of March.

Line 2 started with the model year, represented by the last two digits of the year. The division series and body series must match the first five digits of the car's VIN. The following three letters represented the assembly plant at which the vehicle was built. The final six numbers were the consecutive unit number, which must match the one on the car's VIN.

Line 3 had the car's trim number and color. It included codes for vinyl roof and convertible top colors, if so equipped. These codes are included in each chapter of this text.

The 1967 Camaros had a similar plate, but two additional lines of codes represented options the car came with.

The plate used from 1968 on was slightly redesigned. The time built code was relocated on a third line, which was followed by a modular seat code. The modular seat code was the RPO option code for the type of seat the car was equipped with. For example, an A41 indicated a four-way electric control front seat.

Protect-O-Plates

Original owners of 1966–72 Chevrolet automobiles got a Protect-O-Plate along with the car. It was a metal plate stamped with the owner's name along with lots of coded information, which included major options, VIN, transmission number, axle number, build date and trim codes. The plate's format changed slightly in 1969.

The Protect-O-Plate did not use the RPO codes, but instead used a different coding system. This information is contained in *Publication #11, Chevrolet Models Thru 1975, Chassis and Body Parts Catalog.* Most Chevrolet dealers still use this catalog, or you may want to buy a copy.

Certification labels

Beginning with 1968, all Chevrolet vehicles included a certification label attached on the inside face of the driver's door. The label stated that the vehicle conformed to all applicable safety standards. In addition, the label contained the month and year the vehicle was built along with the car's VIN. This label was revised in the late seventies and eighties, to contain additional information such as vehicle weight ratings.